"Over the past decade, few people have had their finger on the pulse of corporate water challenges as much as Will Sarni. With this book, Sarni once again places himself at the forefront of the next wave of water stewardship. Without embracing digital water, companies will founder in the face of the water-related climate challenges. Conversely, for those who do, digital water will represent an opportunity to scale opportunities through innovation and purpose. *Digital Water* is a great resource for business leaders looking to not just have more effective and efficient water strategies, but to lay a foundation to tackle water challenges in the coming decades."

Alexis Morgan, *Global Water Stewardship Lead at WWF*

"Will has a knack for early leadership on water issues which eventually become mainstream. Not difficult to predict he is doing the same with digital water in this thought-provoking book. Bringing a rather dense topic to life with deep insights and clarity of thought. High risk locations and operations will increasingly understand the value and necessity of digital management & accounting for water . . . this book is a great starting place and early roadmap!"

Andre Fourie, *Global Director, Water Sustainability at Anheuser-Busch InBev*

"*Digital Water* is a must-read book when it comes to understanding how digital technologies will transform the water sector. Sarni provides compelling examples of how organizations are deploying digital tools ranging from IoT to AI to improve the design, situational awareness, control, performance, and engagement of users of water systems. Sarni uses these examples to illuminate the drivers, processes, challenges, and opportunities that come with the use of these digital tools. Digital Water reminds us, as Condoleezza Rice observed, that what once seemed impossible, in retrospect, seems inevitable. Whether you are immersed in the day-to-day issues of digital water or engaged in policy considerations or simply an individual or institutional consumer of water, you will find Sarni's observations and prescriptions to be insightful and incredibly valuable."

Jonathan Copulsky, *Executive Director, Medill Spiegel Research Center, Northwestern University and co-author, The Technology Fallacy and The Transformation Myth*

"Will Sarni has a gift for being in the right place, with the right idea, at the right time. He is the ultimate synthesizer, and this book presents a finely tuned rendering of water's prospective leap into the 21st century. Just as the fountains at Bellagio respond digitally to Handel, Sarni harnesses the information technology trends across industries, and projects them on to water, shining the light on some of the early innovators who are having an impact, and discussing the core principles and obstacles that would shape the future of digital water, from conservation to risk to performance. Will it happen? Will would like you to read and react. So would I. Let's shape the future."

Manu Lall, *Columbia University, Director, Columbia Water Center, Alan & Carol Silberstein Professor of Engineering and Senior Research Scientist, International Research Institute for Climate & Society*

"Will is in the business of saving lives. The status quo is broken and Will Sarni's voice whether live or in print is one that needs to be heard. Will applies his profound industry knowledge as a catalyst for change fueled by his bias for action. Having been personally witness to almost three decades of global water intelligence being central to water professionals, an echo chamber, Will is committed to expanding these boundaries. This work along with many other pieces Will has produced are a prime example of the required thought leadership and education paramount to instigating change and ultimately a reversal of this crumbling status quo."

Alexander Crowell, *Partner, Pureterra*

"The opportunity to rethink water is right here and right now. Start with this book for a clearly articulated path to a resilient future for water where digital technology, investment, policy, and organizations of all shapes and sizes come together. Will Sarni lays the foundation, framing and roadmap for organizations, communities and individuals to transform our approach to the most important substance on earth. Most importantly, see how you can play a part."

Ivan Lalović, *CEO and Founder, Gybe*

"Will Sarni's latest book is a roadmap to help organizations navigate water's digital transformation. He masterfully explains why digital solutions, coupled with a creative destruction mindset, will enable a water abundant future. Readers should come away with a bias for action to help collectively solve the 'wicked' water problem."

Tom Freyberg, *Founder & Director, Atlantean Media*

"If you care about water, this book is for you. Will continues to take complex problems and provide a clear, coherent, and concise assessment of the situation. Examples from other industries shed light on creative solutions. The book leaves you with a better understanding of the digitization of the water sector, how that is breaking down silos, and why it matters. It gives a reason to believe that even though the global water situation seems dire, we have the chance to solve global water problems in our lifetime."

Jon Grant, *Chief Commercial Officer, SENTRY*

"A fast paced and thought-provoking insight into how digital capabilities and the democratization of data are mobilizing collective understanding, action and mobilizing innovation at a scale that hasn't been seen before in the water sector. A taster on what is to come."

Maeve Hall, *Sustainable Manufacturing Water Lead, Unilever*

"An excellent foundational read to get one's mind around the multi-faceted race towards digitalization within the water sector. Will's approach expertly explains how water's roadmap towards digitization rhymes with the successful digitization journeys of other sectors; including the great societal and business benefits that await. A book that is a catalyst to effect positive change."

Matt Rose, *CEO, Apana*

"This book is essential reading for anyone working in the water sector. Will is a globally recognized expert in this field and provides a comprehensive overview of the use of digital tools for water. It combines academic rigor with real-world practicality, giving an in-depth analysis of both the origins, definitions and evolution of digital water techniques, alongside examples and case studies of applications of the technologies. It also enables the reader to compare and contrast digital technologies in water with those from other sectors."

Gaetane Suzenet and Prof Jacob Tompkins OBE,
The European Water Technology Accelerator

"Will Sarni's book delivers the big picture of digital water transformation, which is so much bigger than technology alone. This is what many of us involved in water management have been waiting for: thought leadership, strategy and superb insights."

Dragan Savic, *FREng, Chief Executive Officer*
at KWR Water Research Institute

Digital Water

This book shows how digital technologies are transforming how we locate, manage, treat, distribute, and use water.

Water resources are under stress from over-allocation, increased demand, pollution, climate change, and outdated public policies. Historical approaches to delivering water for human consumption, industrial production, agriculture, power generation, and ecosystems are no longer adequate to meet demands. As a result, we need to vastly improve the efficiency and effectiveness of our public and private sector processes in water management. The author describes recent advances in data acquisition (e.g., satellite imagery, drones, and on-the-ground sensors and smart meters), big data analytics, artificial intelligence, and blockchain, which provide new tools to meet needs in both developing and developed economies. For example, a digital water technology portfolio brings the value of real-time system-wide monitoring – and response – within the capability of water providers of all sizes and sophistication. As such, digital water promises to increase the long-term value of water resource assets while assisting in compliance with regulations and helping respond to the demands of population growth and evolving natural and business ecosystems.

Including many practical examples, the author concludes that digital and smart water technologies will not only better manage water assets but also enable the public sector to provide universal access to safe drinking water, the private sector to continue to grow and ecosystems to thrive.

William Sarni is Founder and CEO, Water Foundry, and Founder and Chairman, WetDATA.org based in Denver, Colorado, USA.

Earthscan Water Text series

Environmental Health Engineering in the Tropics
Water, Sanitation and Disease Control
Sandy Cairncross and Sir Richard Feachem

Flood Risk Management
Global Case Studies of Governance, Policy and Communities
Edited by Edmund Penning-Rowsell and Matilda Becker

Water Ethics
A Values Approach to Solving the Water Crisis
David Groenfeldt

The Water Footprint of Modern Consumer Society
Arjen Y. Hoekstra

Water Governance, Justice and the Right to Water
Edited by Farhana Sultana and Alex Loftus

Digital Water
Enabling a More Resilient, Secure and Equitable Water Future
William Sarni

For more information about this series, please visit: *www.routledge.com/ Earthscan-Water-Text/book-series/ECEWT*

Digital Water

Enabling a More Resilient, Secure and
Equitable Water Future

William Sarni

LONDON AND NEW YORK

from Routledge

First published 2022
by Routledge
2 Park Square, Milton Park, Abingdon, Oxon OX14 4RN

and by Routledge
605 Third Avenue, New York, NY 10158

Routledge is an imprint of the Taylor & Francis Group, an informa business

© 2022 William Sarni

British Library Cataloguing-in-Publication Data
A catalogue record for this book is available from the British Library

Library of Congress Cataloging-in-Publication Data
Names: Sarni, William, author.
Title: Digital water : enabling a more resilient, secure and equitable water future / William Sarni.
Description: Abingdon, Oxon ; New York, NY : Routledge, 2022. | Series: Earthscan water text | Includes bibliographical references and index.
Identifiers: LCCN 2021037737 (print) | LCCN 2021037738 (ebook) | ISBN 9781138343221 (hardback) | ISBN 9781138343238 (paperback) | ISBN 9780429439278 (ebook)
Subjects: LCSH: Water security. | Water-supply—Technological innovations. | Water-supply—Management.
Classification: LCC HD1691 .S2725 2022 (print) | LCC HD1691 (ebook) | DDC 333.91—dc23
LC record available at https://lccn.loc.gov/2021037737
LC ebook record available at https://lccn.loc.gov/2021037738

ISBN: 978-1-138-34322-1 (hbk)
ISBN: 978-1-138-34323-8 (pbk)
ISBN: 978-0-429-43927-8 (ebk)

DOI: 10.4324/9780429439278

For my parents, Michael Sarni and Josephine
Napoli Sarni.

Contents

Figures

Foreword

Digitalization, or digital transformation, is happening everywhere around us; it is affecting the way we live, work, and travel, and it is here to stay! From smartphones and online shopping to digital TV and social media, we are already immersed in the global digital ecosystem. A good question then would be, what is happening with the digital transformation in the water sector? Those of us who are lucky to have access to clean water and wastewater services 24/7 may not be aware of the progress in the digital transformation or even not aware it has been happening as the level of service is already high and we do not have much to complain about. We tend to take utility services as granted and do not pay much attention to them unless there is a breakdown in services they provide, or we are moving house.

The good news is that digital transformation in the water sector is ongoing and has already been around for a long time but has taken off in earnest with the adoption and proliferation of digital computers. For example, computer modeling of various water systems, such as the hydrologic cycle (continuous recycling of water on Earth), the flow in pipe networks (that deliver drinking water to our homes and take away wastewater), and the treatment processes (that purify our drinking water or treat wastewater before it can be released into the environment), have all been the staple of the water practitioner community for decades. The recent explosion of Information and Communications Technologies (ICT) and various sensors (like those in your smartphone or your car) has also accelerated the digitalization process and enables more efficient and reliable monitoring, modeling, and management of water systems. Likewise, those developments have also raised the profile of hydroinformatics (or water informatics), a scientific study that integrates water and information sciences with societal needs, which has also been around for years, even before Professor Michael Abbott coined the name of the discipline in his book "Hydroinformatics: Information Technology and the Aquatic Environment" in 1991. Over the years, the field has grown from purely computational hydraulics into a scientific field in which ICT in its broadest sense is developed for and applied to water management challenges, combining technological, socioeconomic, and environmental perspectives. Digitalization in the water sector can then be seen as the application of ICT tools to improve the design, situational awareness, control, performance, and engagement of users of water systems or subsystems.

However, digitalization is not only about technology, or not just about sensors, sensing data and modelling to enable better management of our water resources. Despite having its origins in computational hydraulics as indicated by Abbott, it does not only concern itself with numerical models, decision support and artificial intelligence (AI) applications in the water sector. The modern field of hydroinformatics also embraces the social dimension of water cycle management (e.g., social needs, concerns, and consequences, including equity, social justice, data privacy, ethics, gender, and legal issues, to name but a few). There are many examples of great technologies that have failed due to the human/social dimensions being neglected. It is not always the most sophisticated or most effective solution that is adopted, but the ones that successfully appeal to users and take them on a path that they feel is important. The success in recent years of some technology companies where some others have failed with, for example, electric cars or smartphones, just emphasizes the fact that technology on its own is not the only deciding factor.

With digitalization, there is inevitably talk about AI and its potential to transform how businesses plan and operate. Over the last two decades, great strides have been made in using AI and machine learning (or data mining) in particular, for better management of water systems. Applications of these technologies being used to detect anomalies in water systems associated with water loss, water quality issues or water demand predictions have now become much more mainstream. Optimization methods, such as genetic algorithms, which are based on mimicking natural evolutionary processes have also gained acceptance in supporting long-term infrastructure investment decisions in the water sector. Serious gaming together with better visualization tools has also attracted interest in the traditionally conservative water sector to allow better engagement with the supply chain and customers.

All these interesting developments have also indicated a possibility of continuous learning through the use of AI and the potential for capturing and storing the knowledge and expertise of an increasingly ageing workforce in the sector. But all of the buzz with AI and "big data" analytics have also raised issues of surveillance-enabled ("big brother") technologies and associated privacy, fundamental rights, ethics and responsibility in technological innovation. One of the key areas that needs addressing by researchers and practitioners involved in developing these new tools is that they are preferably human centric so that AI and machine learning solutions are explainable and acceptable by policymakers, water utility managers/operators, and consumers. Therefore, we are after "augmented intelligence" solutions, rather than super-power AI-enabled machines, that in their purest form aim to have equal or superior intellectual abilities than humans who created them.

Our future is definitely digital, and this book represents an attempt to describe how the world of water will change with it.

Dragan Savić, *Chief Executive Officer at KWR Water Research Institute*
based in the Netherlands and Professor of Hydroinformatics at the
University of Exeter in the UK

Acknowledgments

Thank you to Cassidy White for interviews, research, editing, and pulling this together for submittal, Andrea Fischer for ongoing support, Megan Lampros for guidance and editorial review, and Hillary Mizia, Simone Ballard and Deanna Marie "Drai" Schindler for contributions in research, interviews, manuscript drafts, editing, and graphics.

Thank you to Dragan Savic, PhD, Ivan Lalović, Jonathan Copulsky, Mark Kovscek, Brandon N. Owens, Tim Fleming, Adam Tank, and Clay Kraus for their contributions.

Also, thank you to Piper Stevens for helping me think about the structure of the book, the story arc and most importantly how to scull.

Introduction

Why this book and why now.

I have worked in the world of water for my entire career. Starting as a hydro-geologist working for Geraghty & Miller, projects in my early career involved working with public and private sector enterprises on water supply initiatives in the Northeast United States and Puerto Rico. I learned to log the geology of wells, run well pump tests, analyze test data, and collect water samples with a bailer to send off to the laboratory for analysis. The tools available during this early part of my career are best described as analog.

Now digital technologies are transforming the water sector and our relationship with water in ways that were unimaginable just a few years ago. We can now use satellite data and analytics to map groundwater, real-time water quality, and predict flooding. Digital technologies enable the more efficient and effective management of utility infrastructure and industrial assets and have ushered in an age of smart water in urban and rural homes. Likewise, digital technologies such as IoT devices and artificial intelligence (AI) are increasingly available for the management of utility and industrial assets and support the transition to a digitally enabled human workforce. The water sector, and more broadly water issues and opportunities, have moved the analog world into the digital world. Digital technologies alone have been transformative, but they are also enabling new technologies to address water supply needs (e.g., air moisture capture) and more resilient and sustainable treatment technologies (e.g., local-ized treatment systems for buildings and communities) for improved access to water, sanitation, and hygiene services.

I have always gravitated to new ideas and challenged the status quo. This view of the world is captured by a quote from Peter Diamandis (entrepreneur and founder of X-Prize), "*The day before something is truly a breakthrough, it's a crazy idea.*" The digital transformation of the water sector initially was a crazy idea which has caught my attention for the past several years.

A bit of a disclaimer, there is no way a book can comprehensively capture the current digital water technology landscape. The digital landscape is constantly changing – evolv-ing and expanding to meet new demands and fill emerging market niches. Therefore, this book is less about the details for digital technologies and more about the drivers, processes, challenges, and opportunities that come with the digital transformation. It explores les-sons learned from other sectors on digital transformation and how digital technologies are

DOI: 10.4324/9780429439278-1

an enabler of other innovative and disruptive technologies, business models, and funding platforms.

The book does, however, provide some examples of technologies and the people who are making digital transformation come to life and offers a point of view on where we are headed. Rather than focusing entirely on the water sector, this book will instead provide a view of how digital technologies are transforming humanity's relationship with water. Water is inherently local, yet it broadly encompasses economic, business, social, and spiritual dimensions. It is critical that we consider and understand all of these dimensions if we are to build a more resilient, secure, and equitable future.

As with any of the books I have written on water, my thoughts here have been informed by other innovative thinkers. More often than not, my views are shaped by people in other sectors that are not limited by our traditional view of what is possible for innovation in the water sector. These thinkers embrace the "art of the possible."

Three issues, in particular, have shaped my view on the digital transformation of water: (1) water as a wicked problem, (2) creating abundance and exponential technologies, and (3) the workforce and culture around the digital transformation. Understanding these issues has become central to my view of how we can achieve water security, resiliency, and equity. My hope is that in reading this book and gaining an understanding of the digital water technology landscape you consider these issues and the lessons I've learned from the following three people.

Wicked problems: Tom Higley

Several years ago, Tom Higley, a friend, entrepreneur, and founder of 10.10.10[1] and XGENESIS[2] asked me if water was a "wicked problem." I assumed he was asking whether the "water crisis" (not a term I care for but will use it for simplicity's sake) was a difficult challenge to solve. What I didn't realize was that "wicked problems" actually have a specific definition – one that water fits perfectly within. For me, the most important takeaway from understanding "wicked problems" is that all stakeholders must be engaged in solving these problems as they have unique capabilities and attributes to contribute. For example, entrepreneurs have speed and focus, and the public sector has size and scale (Figure 0.1).

As I read the articles that Tom forwarded, I became even more convinced water challenges should be framed as a "wicked problem." One key article outlining the characteristics of a wicked problem was written by Rittel and Webber in 1973, a summary of which is provided in the following.[3]

- There is no definitive formulation of a wicked problem, that is, even the definition and scope of the problem is contested.
- Wicked problems have no "stopping rule," that is, no definitive solution.
- Solutions to wicked problems are not true-or-false, but good-or-bad in the eyes of stakeholders.

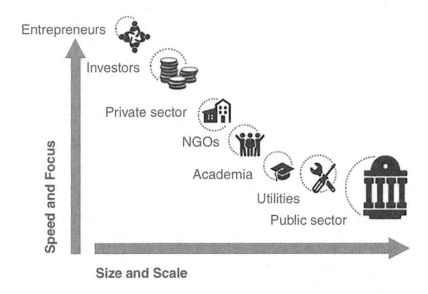

Figure 0.1 Who Solves Wicked Problems

Source: Adapted from T. Higley, 10.10.10 Founders, 2017.

- There is no immediate and no ultimate test of a solution to a wicked problem.
- Every (attempted) solution to a wicked problem is a "one-shot operation"; the results cannot be readily undone, and there is no opportunity to learn by trial-and-error.
- Wicked problems do not have a clear set of potential solutions, nor is there a well-described set of permissible operations to be incorporated into the plan.
- Every wicked problem is essentially unique.
- Every wicked problem can be considered to be a symptom of another problem.
- The existence of a discrepancy representing a wicked problem can be explained in numerous ways.
- The planner has no "right to be wrong" in an experimental sense, that is, there is no public tolerance of initiatives or experiments that fail.

Further to the point that water is a wicked problem, Eddy J. Moors, PhD from the IHE Institute for Water Education framed the issues well.[4]

Water resource management has often been described as a wicked problem, especially because there are no easy solutions. It is wicked because there are unknown dimensions to the related science, with open questions such as: How much water is available? Where is the water coming from? How is this changing in time? What is causing these changes? In addition, there are in almost all cases, multiple stakeholders that deal with the management of water resources. This renders the decision-making difficult and sometimes even impossible. Examples of wicked problems in the water sector are, among others, related to groundwater resources such as the use of the fossil groundwater under a large part of Africa, the fast decline of the groundwater table in the Middle East, but also the strategic management of the groundwater store under "de Veluwe" during prolonged periods of drought. All these issues have, besides a large natural science component, a strong socio-economic component as well.

The University of Arizona Water Resources Research Center also provides insights on why water is a wicked problem from the perspective of water managers.[5]

Reframing water issues as wicked problems will be essential for moving forward to a healthy water future. It begins with considering issues and solutions at the system scale. Water projects can no longer be solely the domain of the water community. Creating a change in one part of the water system will change the entire system, including the dependent social and environmental systems, sometimes in unexpected ways. This will require yet another solution to address that situation. It is an adaptive, iterative exercise. Solutions to wicked problems are never one and done.

Water planners and managers play a central role in mitigating the negative consequences of wicked problems. They will be required to position efforts in new and more desirable directions. This will not be easy, quick, or solitary. It requires methodical, rigorous iteration focused on the system qualities of the problem. The interdisciplinary collaboration that captures a broader knowledge of science, economics, statistics, technology, psychology, politics and more is necessary for effective change.

Managing wicked problems is a new kind of work. It requires changing the questions, managing uncertainty, and creating resilience. It does not solve existing problems but instead drives to a desired future state.

My lesson learned from Higley: wicked problems can arise from complex adaptive systems. My learnings from Tom then are that digital water technologies have a central role to play in solving wicked water problems by democratizing access to data and actionable information to all stakeholders.

Creating abundance: Peter Diamandis

In 2016, I was fortunate to lead a team to create an X-PRIZE Visioneers Team for water to solve the Flint, Michigan lead poisoning tragedy. The team was sponsored by Brita, and we were focused on "democratizing access to safe drinking water and data," with the goal of providing "universal access to safe drinking water, always."[6] While our team was not selected to move forward with building out the prize and securing funding, I became educated on the power of exponential technologies and creating abundance through the work of Peter Diamandis. Peter is the founder and chairman of the X-PRIZE Foundation, cofounder and executive chairman of Singularity University, and coauthor of *The New York Times* bestsellers *Abundance: The Future Is Better Than You Think* and *BOLD: How to Go Big, Create Wealth, and Impact the World.*

Along with others, I again worked with them in 2020 on the development of the Water Abundance X-PRIZE and have closely followed their work and of Diamandis.

There are numerous learnings from Diamandis and X-PRIZE. The most significant for me is the view that we can create abundance by tapping into the power of exponential technologies such as digital solutions. This vision of creating abundance shaped my last two books on water (*Water Stewardship and Business Value* and *Creating 21st Century Abundance through Public Policy Innovation*) as well as this current book on digital water technologies.

My experience with X-PRIZE has also resulted in my focus on exponential technologies, mostly the power of digital technologies but also advances in material sciences which enable innovative air moisture capture technologies used by companies such as SOURCE.[7] To provide some background on exponential technologies and the power of digital technologies, "exponential growth" is when a mathematical function of time's rate of change is proportional to the function's current value. An example of exponential growth in the digital world is Moore's Law, stating that overall computer processing power will double every two years. Deloitte goes further to provide a definition of exponential technology as "innovations progressing at a pace with or exceeding Moore's Law" that "evidence a renaissance of innovation, invention, and discovery. . . [and] have the potential to positively affect billions of lives."[8] More from Deloitte on digital exponential technologies.

> *Exponential technologies are those innovations that continue to advance exponentially, with disruptive economic and lifestyle effects. Moreover, exponential technologies are those whose current price-performance makes it feasible to incorporate them into today's business and social problems in new and previously impossible ways.*
>
> *Examples of exponential technologies you're already seeing in everyday life? Think of artificial intelligence (AI), additive manufacturing, augmented and virtual reality (AR, VR), digital biology and biotech, data science, medical tech, nanotech, robotics, autonomous vehicles, etc.*
>
> *Moreover, and perhaps most importantly, we think that the solutions to some of the world's most urgent problems can be found at the intersection of exponential technologies. When two or more exponential technologies are applied to a serious societal challenge – the chance of creating viable and sustainable solutions increases drastically.*

My lessons from Diamandis: as an exponential technology, digital technologies can contribute to creating abundance. Also, when digital technologies are combined with other exponential technologies such as material science, the chance of creating viable and sustainable solutions increases drastically.

The technology fallacy: Jonathan Copulsky

I worked with Jonathan while we were both at Deloitte Consulting. Jonathan, the Executive Editor of Deloitte Review, and I authored several white papers on sustainability, water strategy, and digital technology value creation. In addition to Jonathan helping improve my writing for the Deloitte Review, his recent book *The Technology Fallacy* provided a wealth of information and insights on how digital transformation doesn't happen without investment in strategy, an enterprise culture of learning and workforce.

Following are a few key takeaways on *The Technology Fallacy* from Sergio Caredda.[9]

The book is not about technology and instead focuses on the human side of digital transformation including organizational structure and culture. The bottom line is that for a digital technology to succeed one must ensure that the company culture is "agile, risk-tolerant and experimental." As framed in the book, *every organization needs to understand its "digital DNA" in order to stop "doing digital" and start "being digital."*

Caredda summarizes the following items representing the gap between companies that are successful in digital transformation and those that are not:

- *communication and decision-making structure, which are too slow in many traditional companies;*
- *mindset and culture, which are normally set for stability; and*
- *digitally mature companies tend to have a flexible and distributed workforce, which leads to having to rethink teams and talent.*

The challenge of digital transformation can ultimately be characterized as a series of "gaps" which relate to the different rates at which people, organizations, and policy responses to technological advances. The varied rates of digital adoption for technology, individuals, businesses adoption, and public policy are illustrated in Figure 0.2 and summarized in the following.

- *The Adoption Gap is the rate at which individuals adopt technologies. One of the challenges vs. the past is that individuals are faster in adopting technologies than companies, and this per se is one of the accelerators of the current disruption.*
- *The Adaptation Gap, which is the rate at which organizations adapt to technology change. One of the issues is . . . the magnitude of investments needed, but to a certain extent there's also the influence of policy and regulations, which lead to the*
- *The Assimilation Gap: institutions, governments, and regulators tend to be even slower in adapting.*

Each of these gaps poses a different challenge for private and public sector enterprises. However, the bottom line is that the ability of the workforce to absorb change is central to a digitally mature organization.

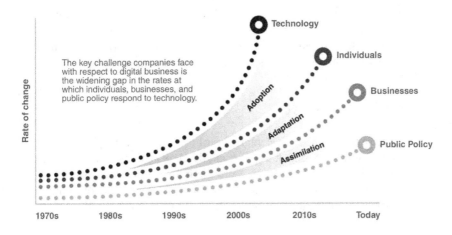

The key challenge companies face with respect to digital business is the widening gap in the rates at which individuals, businesses, and public policy respond to technology.

Figure 0.2 The Varied Rate of Digital Adoption

Source: Adapted from Kane, Phillips, Copulsky, and Andrus (2019).

My lessons from Copulsky: digital transformation is only successful when investment is made in the culture and workforce of an organization. All too often we (myself included) fall in love with a digital technology solution at the expense of the needs of the organization to understand and adopt the technology successfully.

Prateek Joshi, Founder and CEO at Plutoshift

Prateek Joshi, founder and CEO at Plutoshift, recognized early on the power of – and potential for – water data. Water facilities, from utilities to those in the food and beverage or energy sectors, gather a myriad of data regarding how they move, store, use, and clean water. "There's so much data," Prateek says, "I saw an opportunity to put it to use." In founding Plutoshift, Prateek developed an AI software that can take data from multiple, unconnected systems, and reveal valuable insights. The software enables automated performance monitoring for industrial workflows with the aim of saving clients energy and chemical costs by identifying process inefficiencies. "The value we provide is when water operators get to say to their managers 'I can spend X% less and get the same treatment.'"

What sets Plutoshift apart from other technology companies, however, is their focus on their customers.

When you're a company like Coca-Cola or Pepsi, so many companies come and pitch to you, it gets confusing right? They will notice if you just talk about your tech or if you really understand their business. What they want to know is what they get back by investing in Plutoshift. Operators especially want to know if this is just a new project – some piece of homework for them to figure out – or if it will actually make their life easier.

Prateek continues by saying,

One of the key missing elements that I've seen across the product ecosystem is the last mile. It's where you fly in to a facility, sit next to your potential customer, and make sure the information about your product is understood and translated correctly. It's not AI, it's not tools, it's just person to person communication and that's what you do.

Lessons from the expert: Many people are intimidated by AI, but it's not magic, it's simple mathematics automating workflows that were once manual. It's not about replacing people either, it's about automating the parts of work that are repetitive so employees can focus on the things that matter.

Wicked problems, abundance, exponential technologies, and people will be recurring themes in this book as we explore the digital transformation of the water sector.

Notes

1 10.10.10. (2020). https://101010.net/
2 XGenesis (2020). https://xgenesis.io/
3 H. W. J. Rittel & M. M. Webber (1973). Dilemmas in a general theory of planning. *Policy Sciences*, 4, 155–169.
4 E. J. Moors (2017). *Water – Wrestling with wicked problems*. IHE Delft Institute for Water Education. www.un-ihe.org/sites/default/files/inaugural_lecture_eddy_moors_5_october_2017.pdf
5 L. Beutler (2016). *What to do about wicked water problems*. The University of Arizona. https://wrrc.arizona.edu/wicked-water-problems
6 S. Chin (2016). *A Brita XPRIZE to change the world*. The Clorox Company. www.thecloroxcompany.com/blog/brita-xprize-to-change-the-world/
7 SOURCE (2020). www.source.co/commercial/?utm_source=google&utm_medium=cpc&gclid=Cj0KCQiA_qD_BRDiARIsANjZ2LA-ZTm-RUtgT8m4qU-hEhKsgs3tBHFoJMAq2NfmII2ZfF08NlbF1rIaAmLZEALw_wcB
8 Xponential Works (2020). *What is exponential technology?* https://xponentialworks.com/what-is-exponential-technology/
9 S. Caredda (2020). *Book review: The technology fallacy*. https://sergiocaredda.eu/inspiration/books/book-review-the-technology-fallacy/

1   The digital century

As science fiction author Arthur C. Clarke said in his book *Profiles of the Future: An Inquiry into the Limits of the Possible*, "Any sufficiently advanced technology is indistinguishable from magic." While we know digital technologies are far from magic, they have nevertheless had an extraordinary impact, transforming society and businesses in ways that we could not have imagined just a few years ago. As discussed in Chapter 3, digital technology has transformed the transportation, healthcare, education, energy, and entertainment industries. These sectors provide valuable lessons for the digital transformation of the water sector we are now witnessing. While water issues are local, digital technologies provide widespread benefits to all stakeholders engaged in solving wicked water problems. Digital technologies are also enabling innovation in new technologies such as air moisture capture and localized treatment systems.

For perspective, the first workable prototype of the internet came in the late 1960s with the creation of the Advanced Research Projects Agency Network (ARPANET), which was originally funded by the US Department of Defense and allowed multiple computers to communicate on a single network. The technology continued to grow in the 1970s when the Transmission Control Protocol and Internet Protocol (TCP/IP) was adopted which set standards for how data could be transmitted between multiple networks. ARPANET adopted TCP/IP on January 1, 1983, which led to establishing the "network of networks" that became the modern internet in 1990 when computer scientist Tim Berners-Lee invented the World Wide Web.

For many, it is now difficult to imagine a time before the internet.

We are likely not that far from the moment when it will be difficult to imagine a nondigital (analog) water sector and how we manage water supply and demand without digital technologies. Digital technologies provide the ability for real-time measurement of quantity and quality and also enable the adoption of new types of water technologies.

DOI: 10.4324/9780429439278-2

1.1 The Fourth Industrial Revolution

The World Economic Forum (WEF) has been one of the leaders on defining and shaping the impact of the digital revolution. They frame this digital transformation as the Fourth Industrial Revolution (4 IR). The First Industrial Revolution used water and steam power to mechanize production and the Second Industrial Revolution used electric power to create mass production. Most of the world has seen the impact of the Third Industrial Revolution that leveraged electronics and information technology to automate production.

While it is logical to conclude that the 4 IR may just be an extension of the last industrial revolution, it is actually quite distinctive. A primary distinguishing characteristic is how the fusion of technologies "blurs the lines between the physical, digital, and biological worlds."[1] The 4 IR is also characterized by exponential technologies delivering: *velocity, scope, and systems impact.*

For example, we now have billions of people connected by mobile devices, with unprecedented processing power, storage capacity, and access to knowledge. This digital revolution is intertwined with other technology advances in 3D printing, nanotechnology, biotechnology, materials science, and energy storage. Combined, these have clear potential to transform our lives.

Mark Kovscek, Founder and CEO of Conservation Labs

The digital water landscape and technology offerings are constantly evolving as new challenges arise followed by new ways of thinking and the development of innovative solutions. In one example, one individual's – Mark Kovscek – personal experience with an undetected home water leak led to the inspiration for redesigning in-home smart water monitoring systems. Mark has since founded Conservation Labs and launched H2know, their first technology, in May 2020. He recognizes that digital technologies are both a product of their environment and an enabler of environments for further digital adoption. Mark says,

> Digital technologies continue to improve and accelerate. In many ways, H2know could not have existed in a meaningful way just a few years ago. The use of managed services, 5G technology, low-energy components, and numerous cloud technologies makes H2know a reality.

Digital propagates digital. As the water sector embraces digital transformation, it will have an exponential effect, laying the groundwork for future digital solutions and amplifying the resulting conservation and sustainability outcomes.

> *Lessons from the expert*: Don't reinvent the wheel. When developing your digital solution, build on existing data, technologies, and the expertise of others.

As framed by WEF, "the speed of current breakthroughs has no historical precedent and when compared with previous industrial revolutions, the Fourth is evolving at an exponential rather than a linear pace."[2] The advances of the 4 IR are transforming "entire systems of production, management, and governance." Although technology uptake happens at varied rates, 4 IR technologies have been researched in depth by experts at Gartner and BlueTech Research to develop models for describing the timeline, levels and drivers behind the adoption of digital water technologies. WEF has mapped the 4 IR to needs in the water sector to illustrate the current adoption of digital technologies (Figure 1.1).[3]

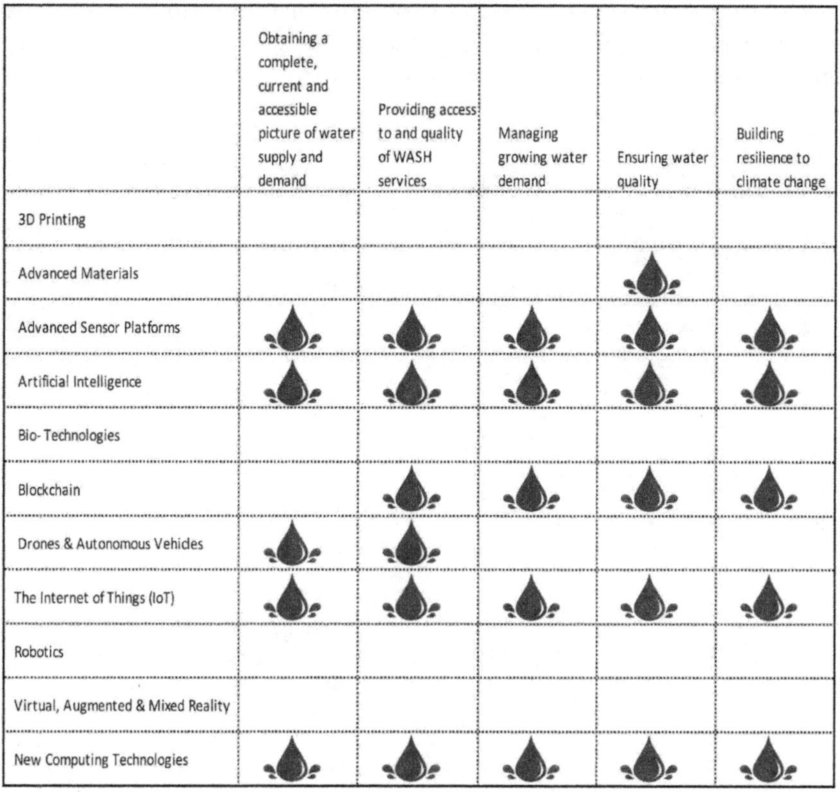

	Obtaining a complete, current and accessible picture of water supply and demand	Providing access to and quality of WASH services	Managing growing water demand	Ensuring water quality	Building resilience to climate change
3D Printing					
Advanced Materials				●	
Advanced Sensor Platforms	●	●	●	●	●
Artificial Intelligence	●	●	●	●	●
Bio-Technologies					
Blockchain		●	●	●	●
Drones & Autonomous Vehicles	●	●			
The Internet of Things (IoT)	●	●	●	●	●
Robotics					
Virtual, Augmented & Mixed Reality					
New Computing Technologies	●	●	●	●	●

WEF_WR129_Harnessing_4IR_Water_Online.pdf

Figure 1.1 The Fourth Industrial Revolution for Water

Source: Adapted from Sarni *et al.* (2018).

1.2 Digital water technology maturity cycle

While the adoption of digital technology over the past few years and in particular during the COVID-19 pandemic have accelerated, it is important to understand the cycles of technology adoption and water technology adoption. A general understanding of the Gartner Hype Cycle and BlueTech Research's view of water technology adoption is provided in the following sections.

These overviews provide a "grain of salt" on the hype of technology adoption.

Gartner Hype Cycle

Digital can be viewed in a similar fashion as other technology innovations. As a result, it is helpful to keep in mind that there is a cycle of introduction, hype, and adoption. The Gartner Hype Cycle is used frequently to frame the maturity or stage of technology adoption.

Researchers at Gartner, an IT consulting firm, have found a common pattern arises as digital technologies mature and are adopted – a pathway they capture in the Gartner Hype Cycle (Figure 1.2).[4]

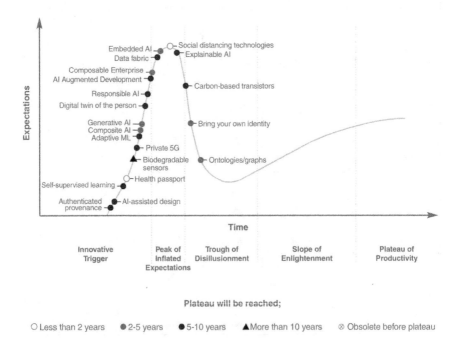

Figure 1.2 The Gartner Hype Cycle

Source: Adapted from Gartner, Inc. (2020).

The Gartner Hype Cycle is divided into a series of five phases that represent distinct levels of maturation, expected market adoption timeline, and public perception of the technology.[5]

- **Innovation Trigger:** Technology breakthrough, early proof-of-concept cases, and media coverage trigger significant public interest. Little in terms of usable available products or proven viability.
- **Peak of Inflated Expectations:** Early publicity and success stories may exaggerate expectations for the technology. Some companies may pursue this technology, but others do not.
- **Trough of Disillusionment:** Interest declines as some experiments and implementations fail to deliver. Many producers drop out or fail and investment only continues if the surviving players can improve their technology to the satisfaction of early adopters.
- **Slope of Enlightenment:** More success stories emerge, added value is realized, and the technology becomes more widely understood. The technology continues to evolve as second- and third-generation products are developed. More enterprises engage through pilot projects, yet conservative companies remain cautious.
- **Plateau of Productivity:** Mainstream adoption takes off. Criteria for assessing provider viability are more clearly defined. The technology's broader market applicability and relevance are understood and paying off for both technology producers and consumers.

The Gartner Hype Cycle can help technology providers understand their technology's current stage of adoption in order to identify next steps for continued maturation and market penetration. In addition, research from Gartner identifies emerging technologies and trends, exploring their current location and expected trajectory along the Hype Cycle. In 2020, Gartner identified five key trends, described in the following, that are expected to significantly impact business and society in the next five to ten years.[6]

- **Composite Architecture:** Businesses are facing a need to shift toward more agile, responsive architectures – a structure that can be provided through a composite architecture design. Made up of "packed business capabilities built of flexible data fabric," a composite architecture enables rapid response to business needs. Modularity and efficiency are core components of composite architecture, enabling businesses to "recompose" themselves as needed (e.g., during a pandemic, recession, and natural disaster). Likewise, the packaged, modular business architecture enables continuous improvement and adaptive innovation (e.g., business becomes more flexible and capable of embracing innovation and adapting to changing business needs/conditions). Digital technologies such as data fabric, 5G, and embedded AI will both be core enablers and drivers of the shift toward composite architecture in business. For successful deployment, however, elements of a composable business must be incorporated across all business functions.

- **Algorithm Trust:** Better data are a critical component of better asset management, operations, more informed decision-making and thus a more sustainable water future. With more data, however, comes an increased risk of data exposure and therefore the need for improved data responsibility. Challenges with fake news, biased AI, and data breaches have led to a rising need for trustworthy data tools and infrastructure. Algorithm trust models are one way to ensure the integrity and security of data, asset information, and personal identities. Other use cases include authentication of assets recorded on blockchain. Whereas blockchain itself is often used for authentication purposes and applied as a tool for transparency, incorrect items added to the blockchain will be continuously verified unless identified by a digital trust algorithm. Experts behind the Gartner Hype Curve expect that, as a demand for blockchain rises, so too will the need for algorithm trust models and more responsible AI.

- **Beyond the Silicon:** Technology is quickly reaching the limits of silicon in regard to data storage, driving the need for innovation in material science in ways that promote development of smaller, faster technologies. New capabilities are emerging in the use of DNA and biochemistry to store data and perform computations. Other emerging technologies include biodegradable sensors and carbon-based transistors.

- **Formative AI:** Formative artificial intelligence technologies are responsive AI that can adapt to dynamic situations or changing technologies. Formative AI can generate novel models to solve specific problems and have vast potential for automation, prediction, and decision-making and creates opportunities in self-supervising data learning.

- **Digital Me:** In the future, digital technologies will increasingly integrate with people and the world around them. Digital twins provide one example in how they utilize sensors to replicate and model physical infrastructure and scenarios. Bidirectional brain–machine interfaces – wearable technologies that enable communication directly between a person's brain and a machine – are another emerging example of the physical technology–biological intersection. Such technologies offer many applications for immersive analytics, account access and payments, and authentication but come with many social concerns over the boundaries and ethics behind connecting technology with the human mind.

BlueTech research

BlueTech Research is a consultancy service based in Ireland that provides analysis of and insights on water sector technologies and technology trends. Research by BlueTech founder and CEO Paul O'Callaghan, PhD has provided a deeper understanding of the stages, drivers, and patterns of technology adoption. Similar to the Gartner Hype Cycle, work by O'Callaghan and colleagues has led to the development of a Water Technology Adoption (WaTA) Model. The model recognizes six

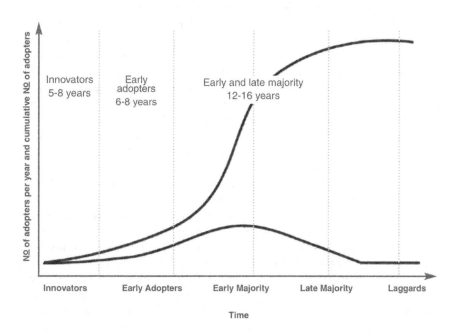

No of adopters per year and cumulative No of adopters

Innovators
5-8 years

Early
adopters
6-8 years

Early and late majority
12-16 years

Innovators   Early Adopters   Early Majority   Late Majority   Laggards

Time

Figure 1.3 The Technology Adoption Curve

Source: Adapted from O'Callaghan *et al.* (2018).

distinct stages of technology maturity: Applied Research, Pilot and Demonstration stages (collectively called the Innovator stages), Early Adopter, Early and Late Majority, and Mature.[7] Figure 1.3 shows the five stages plotted along the WaTA Model (S curve) with reference to a technology diffusion model.

While developing the WaTA curve, O'Callaghan recognized certain technologies were being adopted at faster rates than others. Insights from additional research outlined two primary themes driving technology adoption in the water sector[8]: (1) crisis/needs-driven technology adoption and (2) value-driven technology adoption.

In crisis/needs-driven scenarios, an external driver is found to catalyze the adoption of water technologies. Such drivers include regulation, industry alliances, availability of grants, or immediate environmental or health concerns. Value-driven adoption, on the other hand, occurs as a result of the technology's unique value proposition. Value-driven technologies often generate capital or operational cost savings, have a smaller footprint or a longer lifespan than existing technologies or otherwise have an advantage over alternatives. Researchers noted that, although value-driven technologies followed an adoption timeline typical of the water sector (on the scale of 11–16 years to progress from the Innovator to Early Majority stage), crisis/needs-driven technology adoption occurred up to two times faster and often involved new market entrants.

Whereas external drivers can catalyze crisis/needs-driven adoption as in the case of an emergency health concern or new water quality regulation, O'Callaghan, Adapa and Buisman (2019) warn that technologies can be too early for the risks/crisis that would demand them. For example, a delay in the passing of proposed regulations can be a major setback for technologies that would have been catalyzed by the regulation. That said, a clear value proposition eliminates the need for a crisis to drive adoption and ultimately lowers risk by avoiding dependence on external drivers. Nevertheless, end users are often slower to adopt value-driven technologies without an immediate need or incentive.

In addition to the WaTA model, O'Callaghan describes water technology adoption in terms of "Innovation Diffusion Theory." The theory identifies three categories of technology innovation: (1) radical functionality innovation, (2) discontinuous innovation, and (3) sustaining innovation (Figure 1.4).[9] Innovations considered to have radical functionality accomplish a new task that was previously unknown or impossible. Discontinuous innovations take advantage of new methods or materials to improve on an existing technology or process, and sustaining innovations are those that represent an incremental innovation but do not radically transform the functionality of an existing technology.

Each innovation category comes with a different recommended pathway for building market share and diffusing through the water sector. Sustaining innovations, for one, should be diffused across the industry by incumbents rather than new entrants. Start-ups whose technologies offer radical functionality, however, should aim to be acquired. Discontinuous innovation is similarly led by new entrants and creates market share by threatening existing technologies with obsolescence.

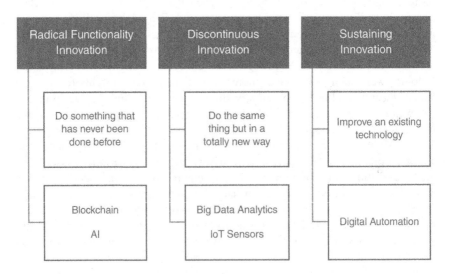

Figure 1.4 Three Types of Innovation

Source: Adapted from O'Callaghan (2020).

In the coming years, the experts at BlueTech Research expect the diffusion of digital technologies to continue. A few highlighted trends include increasing connectivity with IoT sensors in water treatment and distribution systems as well as the need for data analytics and SaaS packages for managing water, sewer, and stormwater systems.[10] Likewise, BlueTech Research expects AI to be increasingly used in all facets of utility operations, from CAPEX planning to remote, intelligent infrastructure assessment.[11] Although many players already provide solutions in these spaces, new technologies are expected to continue emerging.

More so than the design and development of new technologies, however, Paul O'Callaghan emphasizes the need for redesigning systems and facilitating system-level solutions. With new solutions emerging such as IoT Sensors, smart home technologies, digital twins and blockchain, among others, it is at the systems level where digital technologies have their greatest potential.

1.3 Digital and the sustainable development goals

There is an important connection between the 4 IR and our ability to achieve the Sustainable Development Goals (SDGs) including SDG 6: Clean Water and Sanitation (Figure 1.5).[12] WEF has released several publications on the opportunities from digital technologies to achieve the SDGs under the banner of the 4 IR. The overarching thinking on the role of technologies in achieving the SDGs is outlined in the WEF report *Unlocking Technology for the Global Goals and further elaborated in Chapter 6.*

While we can't currently predict how digital technologies will contribute to achieving the SDGs by 2030, there are signs that the impact of COVID-19 has accelerated adoption of digital technologies. While it is perhaps easy to claim that the COVID-19 pandemic has set us back from achieving the SDGs, the pandemic has actually created opportunities to get on track and perhaps even meet our objectives for 2030.

First, where were we prepandemic? As Suma Chakrabarti, chair of the board of the Overseas Development Institute and former president of the European Bank for Reconstruction and Development, made clear in an article from the last fall, "it wasn't like we were on a perfect path [to meet the SDGs] in the first place."[13] In fact, not a single country was on track to complete the SDGs within the decade, given rising rates of inequality, food insecurity, and climate change.

I do agree that we were never on track to begin with. I also believe there is not a simple answer to how much the pandemic has impacted our trajectory in achieving the SDGs.

However, I believe that the adoption of digital technologies *has accelerated during the pandemic and may put us in a stronger position to achieve the SDGs by 2030.* For example, we have rapidly adjusted to working remotely using video conferencing tools that have matured incredibly over the past year, learned to manage teams from afar, orchestrate supply chains, oversee operations, and excel with customer engagement all with digital technologies. This has fostered increased collaboration and productivity beyond measure.

GOAL 6: Ensure availability and sustainable management of water and sanitation for all.
The Pathway to a Sustainable Future.

6.6 Water Related Ecosystems

Protect and restore water-related ecosystems, including mountains, forests, wetlands, rivers, aquifers and lakes

6.1 Safe Drinking Water

Achieve universal and equitable access to safe and affordable drinking water for all

6.2 Sanitization and Hygiene

Achieve access to adequate and equitable sanitation and hygiene for all and end open defecation, paying special attention to the needs of women and girls and those in vulnerable situations

6.5 Integrate Water Resources Management

Implement integrated water resources management at all levels, including through transboundary cooperation as appropriate

6.4 Water Efficiency

Substantially increase water-use efficiency across all sectors and ensure sustainable withdrawals & supply of freshwater to address water scarcity and substantially reduce the number of people suffering from water scarcity

6.3 Water Quality

Improve water quality by reducing pollution, eliminating dumping and minimizing release of hazardous chemicals and materials, halving the proportion of untreated wastewater and substantially increasing recycling & safe reuse globally

6.a   Expand international cooperation and capacity-building support to developing countries in water- and sanitation-related activities and programs, including water harvesting, desalination, water efficiency, wastewater treatment, recycling and reuse technologies

6.b   Support and strengthen the participation of local communities in improving water and sanitation management

Figure 1.5 UN SDG6 Water Sanitation and Hygiene

Source: Adapted from Garkaviy (2015).

More specific to water and its intersection with other resource challenges such as food, digital tools such as the use of satellite data and analytics for real-time water quality monitoring, agriculture management, and water utility asset management (leak and infrastructure failure) have increased in adoption. The ability to more effectively manage water and agricultural production using digital technologies will directly support achieving SDG6 and SDG2 on food security, nutrition, and sustainable agriculture. These kinds of advancements matter.

The pace of innovation during the pandemic has accelerated, and for many public and private sector enterprises, the digital transformation fast track is no longer optional. According to Accenture, "we are seeing three years of digital transformation in three months."[14]

1.4 Digital technologies and water

The world is in a rapid and painful transition from believing that water was plentiful and free (or at the very least inexpensive) to facing the impacts of water scarcity, poor water quality, and the variabilities of hydrologic events from climate change. This realization is slowly unfolding as the public sector faces the realization that policy reform is urgently called for to address human tragedies unfolding in cities such as Flint, Michigan and Cape Town, South Africa. Cape Town is not alone as other global cities face a similar water scarcity crisis: Bangalore, India; São Paulo, Brazil; Beijing, China; Cairo, Egypt; Jakarta, Indonesia; Istanbul, Turkey; Mexico City, Mexico; London, England; Miami, Florida, and Los Angeles, California.

The private sector is also slowly recognizing that water risks impact business continuity and growth along with brand value from their actions in managing water as a shared resource. For those industries that rely upon water for manufacturing (e.g., food, beverage, manufacturing, and semiconductors) access to water is essential for current operations and projected business growth.[15]

What the world is now experiencing can no longer be framed as "normal." The past can no longer be used to predict seasonal weather events and precipitation (e.g., a loss of stationarity). Increasing population growth places demands on the need for water and negatively impacts water quality. As a result, there is a pressing need for: new public policies and business strategies; and innovation in technology, financing, business models, and partnerships to thrive in the 21st century. These new policies, strategies and innovative solutions are only possible with better and accessible data and actionable information. We need to deploy digital solutions (information and communication technologies (ICT)) to enable the more efficient and effective use of water data for public sector business, societal, and ecosystem needs.

The water sector is now in the early but accelerating stages of learning the value of adopting digital technologies to solve water scarcity and quality issues (Figure 1.6).[16] The transformation of the water sector through digital technologies is similar to the experiences had by other sectors. The electricity sector provides perhaps the most relevant example of the benefits of ICT, with dramatic effects that are still unfolding.[17] The move to smart buildings and grids, microgrids, and renewables has been transformative in providing cost-effective electricity in emerging and developed economies. The overall impact and value from adoption of ICT has been well documented through the work of the Global e-Sustainability Initiative (GeSI). GeSI has quantified how the sector through digital technologies has improved energy efficiency and, in turn, reduced greenhouse gas emissions.

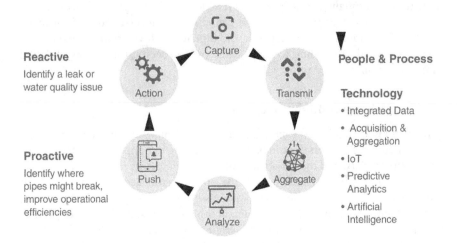

Figure 1.6 Digital Value Creation

Source: Adapted from Raynor and Cotteleer (2015).

Imagine the same transformation in the water sector. Already, digital technologies such as remote sensing can provide vastly improved predictions of droughts and flooding, real-time monitoring of water quantity and quality within watersheds, improved water utility asset management, off-grid and localized solutions coupled with real-time water quality monitoring and "frictionless" water trading platforms. Digital technologies such as connected devices (IoT), predictive analytics, and artificial intelligence are also emerging as powerful tools in achieving sustainable, resilient, and equitable access to water.

Most recently, GeSI published *System Transformation: How Digital Solutions Will Drive Progress Towards the Sustainable Development Goals*. The summary report includes a discussion of how ICT is playing the central role in achieving Sustainable Development Goal 6 through water efficiency and increasing access to water through "smart water management" – smart pipes, smart meters, soil sensors, remote irrigation management systems, consumption control applications, and e-billing.[18]

1.5 What's next?

A recent issue of *The Economist* featured an article titled "The Plague Year: The Year When Everything Changed." Few will be surprised by this title. However, aside from the COVID-19 reflections, the article provides insights about lessons learned from several events from the early part of the last century. The article explains that "The horrors of the first world war and the 'Spanish Flu' were followed by the Roaring Twenties, which can be characterized by risk-taking social, industrial and artistic novelty."[19]

In 1920, US presidential candidate Warren Harding built his campaign around "normalcy." What unfolded was not a return to normal. According to the economist, the survivors of the Spanish Flu and the First World War left survivors with "an appetite to live the 1920s at speed."

I am willing to place a bet that our view of water, including the more traditional view of the water sector – think utilities, solutions providers, NGOs – will not return to normal. And, frankly, we shouldn't go back. The water sector from a technology, business model, and funding perspective will be transformed, driven by lessons learned from the pandemic but also due to the natural rhythm of "creative destruction."

I believe this is the year where creative destruction will transform the water sector and our view of water. In the early 20th century, the economist Joseph Schumpeter described the dynamic pattern in which innovative entrepreneurs unseat established firms through a process he called "creative destruction."

According to Schumpeter, and discussed in detail in "The Prophet of Innovation: Joseph Schumpeter and Creative Destruction," it is the entrepreneur who not only creates invention but also creates competition from a new commodity, new technology, new source of supply, and a new type of organization.[20] The entrepreneur creates competition, "which commands a decisive cost or quality advantage and which strikes not at the margins of the profits and the outputs of the existing firms but at their foundations and their very lives." This innovation propels the economy with "gales of creative destruction," which "incessantly revolutionizes the economic structure from within, incessantly destroying the old one, incessantly creating a new one."

Schumpeter's view of creative destruction was applied to the emerging trend of sustainability in 1999 by Stuart L. Hart and Mark B. Milstein in their article, "Global Sustainability and the Creative Destruction of Industries." It was this article that first got me curious about the cycles of creative destruction and its relevance to the water sector.

For me, the key point from Hart and Milstein is that with technological innovation, there is a dramatic transformation in institutions and society.[21] The technology innovation – and, in turn, transformation in institutions and in society – creates profound challenges to incumbent businesses. Historically, these incumbents (the installed base) "have not been successful in building the capabilities needed to secure a position in the new competitive landscape."

What does creative destruction and disruptive innovation mean for the water sector? I believe it will, in general, be the democratization of water. It will be an "end run" around the public sector, infrastructure and traditional financing of innovations to deliver universal and equitable access to safe drinking water, sanitation, and hygiene.

The creative destruction of water will include real-time and actionable information on water quantity and quality and increased access to capital to scale innovative solutions such as air moisture capture, decentralizing water treatment and reuse systems at the residential and community scale.

The water sector is poised to undergo a "gale of creative destruction" to a large degree instigated by the pandemic. The accelerated transition to using digital technologies is also an enabling tool, in addition to providing readily accessible actionable information to the general population. I believe we are now entering the roaring 20s for water.

1.6 A Word (or two) of caution

I am an avid believer that the application of digital technologies is the most important trend in the water sector. However, innovation is lumpy, and outcomes can be unexpected. We need to be mindful that immediate gains in productivity didn't come quickly from the birth of the ICT sector and the internet. In the United States, the Labor Department statistics in the late 1990s signaled an increase in productivity from the ICT sector. However, this productivity was not sustained and stalled around 2005. Several factors could account for this including the 2008 recession.[22]

Time will tell how the digital transformation of the water sector will unfold and if our predictions of value creation and productivity gains are realized and over what period of time. The central issue with digital transformation of a company, public sector, industry, etc., are people. It is wise to keep this issue at the forefront when we consider disruptive innovation such as digital technologies.

Notes

1 K. Schwab (2016). *The fourth industrial revolution: What it means, how to respond.* World Economic Forum. www.weforum.org/agenda/2016/01/the-fourth-industrial-revolution-what-it-means-and-how-to-respond/

2 Ibid.

3 W. Sarni, C. Stinson, A. Mung, & B. Garcia (2018, September). *Harnessing the fourth industrial revolution for water.* Fourth Industrial Revolution for the Earth Series. www.weforum.org/reports/harnessing-the-fourth-industrial-revolution-for-water

4 K. Panetta (2020). 5 Trends drive the Gartner hype cycle for emerging technologies, 2020. *Gartner.* www.gartner.com/smarterwithgartner/5-trends-drive-the-gartner-hype-cycle-for-emerging-technologies-2020/

5 Gartner Hype Cycle (2021). *Gartner.* www.gartner.com/en/research/methodologies/gartner-hype-cycle

6 K. Panetta (2020). 5 Trends drive the Gartner hype cycle for emerging technologies, 2020. *Gartner.* www.gartner.com/smarterwithgartner/5-trends-drive-the-gartner-hype-cycle-for-emerging-technologies-2020/

7 P. O'Callaghan, G. Daigger, L. Adapa, & C. Buisman (2018). Development and application of a model to study water technology adoption. *Water Environment Research*, 90(6), 563–574.

8 P. O'Callaghan, L. M. Adapa, & C. Buisman (2019). Analysis of adoption rates for needs driven versus value driven innovation water technologies. *Water Environment Research*, 91, 144–156.

9 P. O'Callaghan (2020). *Dynamics of water innovation: Insights into the rate of adoption, diffusion and success of innovative water technologies globally.* Wageningen University.

10 R. Retief (2016). The internet of water: Impacts and opportunities in the water technology market. *BlueTech Research.*

11 G. Symmonds (2020). AI and decision support tools. *BlueTech Research.*

12 E. Garkaviy (2015). *Pathway to a sustainable future – Infographic.* Hope Spring. www.hopespring.org.uk/water-and-sanitation-the-pathway-to-a-sustainable-future-infographic/

13 A. Lieberman (2020). How off track are the SDGs, exactly? We don't know, but it might not matter. *Devex.* www.devex.com/news/how-off-track-are-the-sdgs-exactly-we-don-t-know-but-it-might-not-matter-98125

14 V. Afshar (2020). Accenture, acceleration: Three years of transformation in three months. *ZDNet.* www.zdnet.com/article/accenture-we-are-seeing-three-years-of-digital-transformation-in-three-months/

15 CDP (2020). *Water.* www.cdp.net/en/water

16 M. E. Raynor & M. J. Cotteleer (2015). The more things change: Value creation, value capture, and the internet of things. *Deloitte Review,* 17, 50–65.

17 S. Erlinghagen & J. Markard (2012). Smart grids and the transformation of the electricity sector: ICT firms as potential catalysts for sectoral change. *Energy Policy,* 51, 895–906.

18 GeSI (n.d.). *System transformation.* https://gesi.org/research/systemtransformation.

19 The Economist (2020, December 16). *The year when everything changed.* www.economist.com/leaders/2020/12/19/the-year-when-everything-changed

20 T. K. McCraw (2009). *Prophet of innovation.* Harvard University Press.

21 S. L. Hart & M. B. Milstein (1999). Global sustainability and the creative destruction of industries. *MIT Sloan Management Review.* https://sloanreview.mit.edu/article/global-sustainability-and-the-creative-destruction-of-industries/

22 J. Cassidy (2013, April 1). What happened to the internet productivity miracle. *The New Yorker.* www.newyorker.com/news/john-cassidy/what-happened-to-the-internet-productivity-miracle

2 "Wicked problem" of water

Why business as usual is failing us

In her book *Beyond Disruption: Changing the Rules in the Marketplace*, Jean-Marie Dru notes that, *"Disruption is rooted in life itself . . . Life's essence lies in accidents and interruptions, in conflict and tension."* As conflicts arise over the quantity, quality, reliability, and accessibility of water, large-scale disruption will be essential, but what has to change in the world of water? Change is not just the water sector but humanity's relationship with water. In my view, everything needs to change. Our current water roadmap for water: extract, transport, treat, and discharge needs to be replaced with diversifying sources of water including air moisture capture, reuse, conservation, and utilities as "factories" for nutrients, energy, and treated water (potable and nonpotable). Figure 2.1 illustrates this roadmap transition.

Unfortunately, this change is not easy to say the least. An invaluable way to view why changing the status quo is viewing water as a wicked problem. A brief recap of wicked problems and how they drive innovation is critical to understanding how we execute on the 21st-century water roadmap.

The relevance of wicked problems was immediately apparent as all things related to water have many, if not all, of the attributes of a wicked problem. From my personal perspective and experience, it is helpful to frame water as a wicked problem to acknowledge not only the challenges but also the opportunities to solve these problems.

Building upon the earlier overview of wicked problems, here are a few more insights on wicked problems and how they create opportunities for stakeholders such as entrepreneurs, investors, and others. Adapted from Tom Higley:

- *Wicked problems arise within, and are held in place by, complex adaptive systems.*
- *Wicked problems result from the mismatch between how real-world systems work and how we think they work.*
- *Systems thinking attempts to resolve this mismatch.*

While I often talk about water as a wicked problem, I am also learning more about complex adaptive systems which can result in wicked problems. Complex adaptive systems are characterized where no one stakeholder has direct

DOI: 10.4324/9780429439278-3

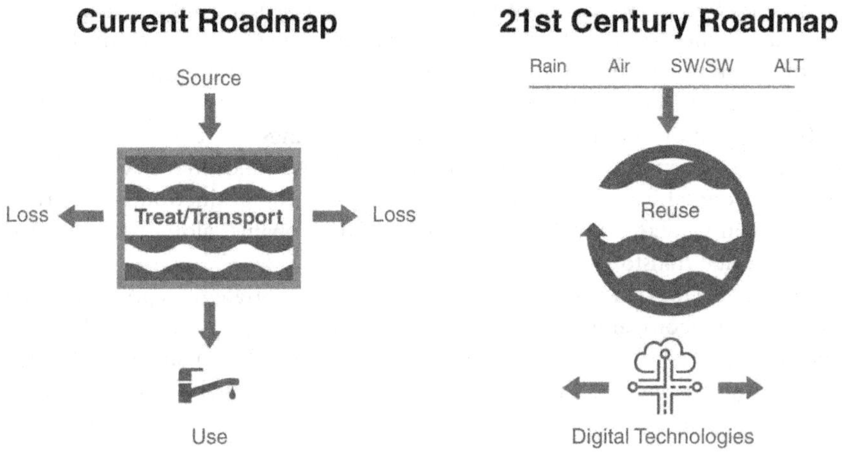

Current Roadmap

21st Century Roadmap

Figure 2.1 What Has to Change: A 21st-Century Roadmap
Source: Adapted from Lux Research Water Intelligence (2008).

control, stakeholders act as independent agents, there are complex feedback loops, and the systems are influenced by cultures and norms.

The ability to influence and change complex adaptive systems and resulting wicked problems is almost the result of a change agent who is an "outsider" to the system. The change agent is not dependent on the system, has the ability to challenge the system, can either work with the system to change it or can disrupt it and often is funded (e.g., a start-up) from sources outside the system.

Why is it important to understand complex adaptive systems and wicked problems in the world of water? *Every wicked problem is an opportunity to engage everyone to solve wicked problems.*

If we can appreciate the complexity of water challenges and, in turn, identify and support change agents, we have an opportunity to solve wicked water problems. The solutions will come in innovative technologies (e.g., digital), business models, partnerships, investing, and public policies. The process for turning wicked problems into opportunities starts with identifying the change agents and stakeholders and requires answering; how do they act, who benefits, who loses and what are the feedback loops, cycles, and cycle times? Other key questions on the journey to identifying opportunities are: who are the people and groups of people who experience these wicked problems, those people who make commitments to disrupt the system and those people who can gain access to the means of coopting the system or forcing change via technological, financial, or political means (or via all three).

One of the key questions is why the status quo can't effectively address wicked problems? I believe one of the answers is that there are few incentives to change as we are complacent with "good enough" instead of focusing on how to do better. Also, wicked problems are market opportunities but lack an obvious business model. It is the outsider as a change agent that can develop innovative business models often when insiders can't see the opportunities.

We urgently need to disrupt the status quo in water.

Not just innovation but true disruption. This brings me to a brief discussion on the distinction between innovation and disruptive innovation. When thinking about digital water technology transformation, the distinction is important in how you view the numerous and established companies along with the process of digital transformation.

Different types of innovation require different strategic approaches for industry and companies.[1] According to the work of Christensen, Raynor, and McDonald, "disruption" describes a process whereby a smaller company with fewer resources is able to successfully challenge established incumbent businesses.

> As incumbents focus on improving their products and services for their most demanding (and usually most profitable) customers, they exceed the needs of some segments and ignore the needs of others. Entrants that prove disruptive begin by successfully targeting those overlooked segments, gaining a foothold by delivering more-suitable functionality – frequently at a lower price. Incumbents, chasing higher profitability in more-demanding segments, tend not to respond vigorously. Entrants then move upmarket, delivering the performance that incumbents' mainstream customers require while preserving the advantages that drove their early success.

When mainstream customers start adopting the entrants' offerings in volume, disruption has occurred.

Other key points:

1. **Disruption is a process**
    The term "disruptive innovation" is misleading when it is used to refer to a product or service at one fixed point, rather than to the evolution of that product or service over time.

    Almost every innovation, disruptive or not, begins life as a small-scale experiment. Disrupters tend to focus on getting the business model, rather than merely the product, just right. Almost all water technologies start as a "small-scale experiment" aka, a pilot. The water sector has an abundance of pilots.

2. **Disrupters often build business models that are very different from those of incumbents** For the water sector, this is "water as a service" as opposed to selling technology and capital-intensive projects. WaaS includes performance contracting but also outsourcing water treatment operations and paying for the treatment.

3. **Some disruptive innovations succeed; some don't**

   A third common mistake is to focus on the results achieved – to claim that a company is disruptive by virtue of its success. But success is not built into the definition of disruption: Not every disruptive path leads to success and not every success follows a disruptive path. The water sector is littered with disruptive innovations that fail for a range of reasons: the team, the business model, lack of adequate funding, etc.

4. **The mantra "Disrupt or be disrupted" can misguide us**

   Incumbent companies do need to respond to disruption if it's occurring, but they should not overreact by dismantling a still-profitable business. Instead, they should continue to strengthen relationships with core customers by investing in sustaining innovations. In addition, they can create a new division focused solely on the growth opportunities that arise from the disruption. The success of this new enterprise depends in large part on keeping it separate from the core business. That means that for some time, incumbents will find themselves managing two very different operations – skunkworks.[2]

   Christensen, Raynor, and McDonald's research indicate that using disruptive theory helps in making more accurate predictions of which fledgling businesses will succeed. This is invaluable when viewing the digital transformation of the water sector.

Notes

1 C. M. Christensen, M. E. Raynor, & R. McDonald (2015). What is disruptive innovation? *Harvard Business Review*. https://hbr.org/2015/12/what-is-disruptive-innovation

2 W. Sarni (2020). *For water innovation to fly, we need a skunk works*. Aqua Tech. www.aquatechtrade.com/news/industrial-water/will-sarni-we-need-a-skunk-works/

3 The digital water opportunity

The experience of other sectors

The digital transformation of the water sector must not be an ad hoc application of new technologies, rather than the intentional, organized implementation of strategy and solutions. We must keep in mind that, as George Westerman of the MIT Sloan Initiative on the Digital Economy puts it, "When digital transformation is done right, it's like a caterpillar turning into a butterfly, but when done wrong, all you have is a really fast caterpillar." In this regard, the water sector can turn to lessons from other sectors on how to approach and what to expect from digital transformation. For example, the transportation, healthcare, education, and entertainment sectors have all adopted digital technologies to varying degrees. The experiences of innovators and other stakeholders in these sectors provide valuable insights on challenges and opportunities in digital technology innovation and transformation.

3.1 Leapfrogging, coupling, and decoupling

Humans are masters of adaptation. From the Romans to the Aztecs to modern-day, we have altered our surroundings to create environments and infrastructure that allow human life to thrive. Water, sanitation, energy distribution, telecommunications, transportation, etc. from concept to widespread implementation, these key modern infrastructures have taken hundreds of years to occur, and then only in well-populated urban areas or nations with the financial means. Implementation and maintenance still take decades of planning, with dire consequences when proper time and resources are not allocated. One only needs to look to the ongoing water crisis in Flint, Michigan, or the rolling blackouts in 2013 that left 150 million people without power in Bangladesh to be reminded of that.

When it comes to water, many infrastructures around the world are either failing or nonexistent. According to the 2017 American Society of Civil Engineers (ASCE) Report Card for Infrastructure, the water infrastructure for the United States rates a D on a scale of A–F. In Africa, only 58 percent of the population has access to clean drinking water, and an estimated annual investment of $15 billion would be required to reach the remaining 42 percent.[1] The time and money required to upkeep aging systems that were designed for another era seems insurmountable.

DOI: 10.4324/9780429439278-4

But what if we didn't have to think about updating or creating infrastructure in the same way we did when it was new? Leapfrogging, the concept of jumping over one or more generations of previous infrastructure to arrive at a new infrastructure technology suited for today's world is happening all around the globe and across many sectors. Perhaps the best-known example is the move from no or extremely poor telecommunications in Africa, India, Asia, and South America, to a highly utilized network of cell phones and mobile devices. As one person put it, "India, China and Africa are all building out their communications infrastructure on the back of the cell phone and not the copper wire."[2]

The rapid arrival of mobile phones has led to leapfrogging in the finance, education, energy, and even healthcare sectors. Kenya is an example of this multifaceted leapfrogging, where two innovative phone-based programs, M-PESA that provides nonbank-centered monetary transactions and M-TIBA that connects people to healthcare and health insurance, have helped the country see a reduction in poverty, an increase in financial transactions, and will hopefully support a healthier citizenry.[3]

Many developing countries are forgoing fossil fuels, which come with high environmental and social costs, and in some cases, the centralized distribution grid required by a singular large power source. From localized solar gardens and individual solar installations at affordable prices to wind power, these technologies simultaneously leapfrog over outdated fuel sources and infrastructure while also decoupling from traditional environmental impacts caused by energy generation. The result is equitable access to energy, decreased pollution-related deaths, increased job growth,[4] and a positive impact on the ecosystems usually affected by fossil fuel extraction. These increased benefits to society leak from one sector to another, helping to both spur innovation and create a cleaner world.

In fact, some of the most effective work being done with infrastructure technology is reliant on the power of entities, even those that may traditionally be at odds, working together to create new, powerful solutions. As Joshua Sperling of the National Renewable Energy Laboratory explains, coupling involves application

> across interfaces of rural to urban, district-city-national-to global scale, to social-ecological-infrastructural systems, or even across public-private-research sectors – that together can effectively enable new synergies between industries, technologies, infrastructure, or policy trajectories that maximize economic prosperity (or productivity) while enhancing environmental, resource, and service sustainability (or resilience), respectively.[5]

These solution-oriented partnerships bring far greater resources to the research and implementation of meaningful actions than any party working in isolation.

The Africa Renewable Energy Initiative is one such coupling worth examining. With a goal to achieve at least 10 GW of new and additional renewable energy generation capacity by 2020 and mobilize the African potential to generate at least 300 GW by 2030, it is a robust initiative with necessarily aggressive goals.

3.2 Sector transformation

As the water sector undergoes its digital makeover, it is important to remember that the water sector is not alone in its transformation. In fact, there is much to be learned from the experiences of other sectors. The telecommunications, personal mobility, healthcare, and energy sectors have all experienced drastic changes in their operations, business strategy, and customer service due to the implementation of digital technologies.

Telecommunications

The telecommunications (telecom) industry has been both influenced by, and a primary enabler of, the digital revolution. As the provider of communications infrastructure, online networks, and cloud services, telecom provides the building blocks for other sectors' digitalization through access, interconnectivity, and applications.[6] As consumer interests have evolved, with customers now requiring rapid, on demand services, and new applications emerging to bypass traditional mediums for communication, telecom has seen a decline in revenue.[7] However, the new expectations, in combination with new business models and technologies, have provided an opportunity for the telecom industry to reinvent itself.

The telecom industry has always been adaptive to changing trends and emerging technologies; however, the transformation of the sector took on a new direction with the appearance of text, call, and video offerings such as WhatsApp, Facebook Messenger, and Skype.[8] In addition, consumers have been demanding faster networks, broader access, and on-demand services. Customers are also increasingly interested in video content including YouTube channels, social media streaming, and over-the-top sites such as Netflix and Hulu. Such changes in market demands are leading providers to turn to new business models including bundling alternatives to full services and subscription models. Telecom has thus found itself at the heart of digital transformation, directly influencing the industry's need to improve current technology and develop new technologies to supply the cloud, streaming, and communications services customers now prefer.

As telecom evolves to meet the changing market demands, new technologies and services are emerging. Fixed broadband services, fixed mobile convergence, cloud technologies, mobile finance services, and various other applications are rising to fill the gap left by declining traditional services (e.g., fixed lines, cable television, and mobile communications).[9] Service providers are finding they need to adapt continuously to changes in technology and consumer demands, enhancing existing offerings and introducing new products and services to meet consumers' evolving needs.[10] Telecom has also learned to capitalize on the large volumes of IoT data already in their possession, using insights from IoT data to optimize bandwidth and coverage, reduce dropped calls, and boost download speeds, minimizing the lost revenue due to service disruptions.[11]

Such advancements are leading to the digitalization of other sectors by providing the technology and services other industries require (e.g., equipment and network services that enable IoT, big data acquisition, and cloud). By focusing on helping users securely connect to their applications and manage their data, the telecom industry is embracing the digital era. Evolving from expensive, difficult-to-manage network elements to virtualized communications and cloud infrastructure that can be managed autonomously and at a low cost, the telecommunications industry provides one example of the digital transformation experienced by other sectors.

Personalized mobility

The transportation industry has been quick to incorporate digital technologies with computer systems, GPS, and sensors already long incorporated into most traditional modes of transportation (e.g., cars, planes, trains, and ships). As the digital era unfolds, however, modes and methods for mobility have seen drastic changes alongside the type and frequency of transportation. New realities are becoming the norm. Virtual activities are becoming an alternative to physical activities (e.g., telecommuting, online shopping, online health services, and education), reducing some types of personal transportation. The rise of digital services and change in social behavior has been a driver for change in the transportation sector – changes that are only predicted to continue in the coming decades.

With the world at their fingertips in the form of smartphones and instant information and entertainment, consumers have begun demanding that same instantaneity in other aspects of their lives. In addition, consumers only want to pay for exactly what they need. In that sense, traditional ownership of personal transportation vessels is becoming a thing of the past. Instead, Mobility as a Service (MaaS) is enabling a transition to subscription services, wherein a subscriber has access to a variety of services that can be purchased as needed (Figure 3.1).[12] MaaS services can include bikes, scooters, cars, and individual ride services. Rather than subscribing to MaaS, cities around the world also offer consumers the ability to rent individual services (e.g., one time use of a bike or scooter), with digital technologies automating locking/unlocking, collecting payment, and identifying bike/scooter locations. Similar rental services are also becoming available for private car use.

Subscription and rental services, however, are just one component of the transportation industry's digital transformation. Dynamic carpooling and ridesharing services are largely changing how people travel. Using an app, vehicles are "e-hailed," with the driver, route, payment, and other passenger pickups (for carpooling services) all being managed electronically. With the rise of ride-hailing services, there is less of a need for individuals to invest in ownership of vehicles, but also increased flexibility is provided to consumers.

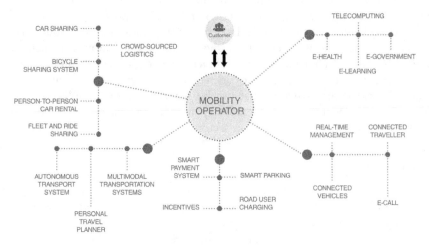

Figure 3.1 Digital Transformation of the Transportation Sector

Source: Adapted from Garcia *et al.* (2019).

Whereas the digital revolution has already largely influenced the transportation sector, there is still much potential for digital technology incorporation and further industry transformation. For example, with the IoT, sensors, and AI technologies now available, autonomous vehicles are nearly a reality for individual travel. One study estimates autonomous vehicles will gain 50 percent market penetration by 2050.[13] Already at play in the trucking industry, autonomous vehicles can utilize AI for intelligent routing to avoid congestion, advanced sensors to maximize safety, and big data and IoT to measure vehicle performance, enabling predictive maintenance.[14] Another potential for the transportation sector lies in IoT and its ability to connect cars, street lights, and traffic signals, enabling smart transportation systems and improving the reliability, efficiency, cost-effectiveness, and safety of travel as well as enhancing asset management and planning.

Digital technologies have already played a major role in the transformation of personalized mobility, revolutionizing when and how people travel, and with intelligent networks and increased automation entering the sector, consumers will continue to reap the safety, efficiency, and cost benefits of digitalization in the transportation industry.

Healthcare

Digital technologies have allowed for revolutionary transformations in the healthcare industry (both in primary care and health insurance). With the health industry's notorious high cost of care in the United States, the drive for digitalization in healthcare stems from the immediate need for control and efficiency optimization alongside longer-term goals such as increased precision, fewer errors, and better outcomes.[15] In addition, the healthcare industry's ongoing shift from a disease treatment to a health management focus requires an ability to monitor conditions in order to predict and prevent adverse health effects. With the advance of digital technologies, new and improved methods are emerging for treatment, monitoring, and general patient interactions (e.g., consultations and health service payment) both within and outside of the hospital environment (Figure 3.2).[16]

Perhaps the most impactful effect of digitalization on the healthcare industry is the role of digital technologies in diagnostics and treatment. Diagnostic and treatment errors are still relatively common, and new technologies can improve treatment precision and decrease the probability of error in disease identification.[17] AI systems have the capacity to analyze thousands of pathology images to provide highly accurate diagnoses, helping radiologists where details may be missed by the human eye. AI can then aid in developing personalized treatments tailored to the genetic makeup and lifestyle of a patient.[18]

There are many other ways in which AI technologies are transforming the healthcare industry as well. AI algorithms are being used to analyze in-patient data (e.g., monitoring vitals and other conditions), and robots are being developed to fetch and restock supplies,[19] freeing hospital personnel from routine patient visits.[20] Similarly, pharmaceutical and biotechnology companies are using machine learning algorithms to shorten drug development life cycles.[21] Other AI applications range from chatbots and virtual health assistants that help with customer service to robots simulating and in some cases performing minor surgeries and biopsies through remote control.[22] For example, as early as 2015, surgeons at the Florida Hospital Nicholson Center were using surgical robot systems performed on simulators in Texas, over 1,200 miles away.[23]

In addition to AI, other big data analytics tools are aiding in lowering the rate of medication errors through patient record analysis. Big data analytics can identify patients who frequently visit emergency rooms and develop preventative care plans for them. Accurate staffing is also enabled by digital technologies that estimate admission rates. Such predictive technologies help pharmaceutical companies better understand the market and can estimate which illnesses or diseases may experience outbreaks in the future.

VR technologies are making headway in the healthcare industry through their abilities to treat pain, anxiety, PTSD, and strokes as an alternative or supplement to drug treatments. Likewise, VR headsets can motivate wearers to exercise and help autistic children learn to navigate the world. VR is also being used to train doctors and residents using scenario simulations and in planning complicated surgeries.

Figure 3.2 Digital Transformation of the Healthcare Sector

Source: Adapted from "The impact of digital transformation in healthcare" (2021).

Sensors and the IoT are also leaving their mark on the healthcare industry. Not only can devices and monitors now communicate among each other in a hospital setting, but also wearable sensors are transforming patient monitoring and treatment both prehospital and posthospital visits. Wearable devices (e.g., heart rate sensors, exercise trackers, sweat meters, and oximeters) can monitor high-risk patients; determine the likelihood of a health event; provide incentives for healthy, active behavior; and monitor patients poststroke, heart attack, or other extreme health event.[24][25] For example, in 2016, a pilot program by the National University of Singapore used an IoT-based tele-rehabilitation program for stroke patients. Patients were issued wearable devices to monitor their condition, and care providers guided patients through rehabilitation exercises remotely using tablets. The program ensured regular therapy without the cost to the patient of traveling to a care center, at the same time saving the hospital time and money by reducing house calls.[26]

The hospital environment and healthcare industry have long been an ever-growing sea of data, patient files, and health and insurance forms. Blockchain technologies, however, are improving the accuracy of data records, preventing data breaches, and cutting costs for hospitals. By providing the ability to record transactions, consolidate patient information, and detect duplicate patient records, blockchain is streamlining the healthcare and health insurance industries, simplifying and securing processes for both patients and providers.

Education

The digitization of education comes with a sweeping positive impact on society. By allowing a host of degrees, from high school to PhD, and educational experiences beyond the traditional classroom to a wider range of students, obtaining a degree has never been more attainable for more people. While the use of internet-based learning, such as online courses and the use of a virtual campus, has been available since the 1990s,[27] present-day digitized education solutions involve far greater reach and creativity. By jumping past the traditional classroom, the education sector has been able to bring new opportunities quickly and effectively to the billions of traditional and nontraditional students alike.

In 2019, India launched a major initiative to digitize state-run schools and colleges to further support a model of "flipped classrooms,"[28] a pedagogical approach that exposes the students to a concept outside of class and then uses in-class time to process and personalize the learning. With easier access to information at any time of day or night that is best for the student, digital resources have the ability to increase both the effectiveness of education and reach a wider range of Indian students. Whether flipping the classroom or not, the simple act of creating a system where a greater number of students have access to the same experience and exposure has the potential to bring an equity to education that has been missing for many.

To be clear, simply handing a student, a digital device does mean their learning will be better, easier, or more equitable. Perhaps one of the greatest lessons we can take from the education sector's exploration into digital tools is that digitization is most effective when paired with a larger plan. Educational goals have been directly linked to the incorporation of digital technologies in the classroom, with a clear outline for how each technology will contribute to students' achievement of academic endpoints.[29] It is important to recognize, however, especially in light of the COVID-19 pandemic and the respective shift to online learning that extensive inequities still exist around access to digital technology and connectivity. Moving forward, increased connectivity will be critical for enabling equitable access to digital technologies in the education, water, and other sectors.

Renewable energy

Digital technologies are revolutionizing the way energy and fuel is explored, collected, generated, stored, distributed, and consumed across the energy sector (Figure 3.3).[30] Perhaps the most transformative impact of digitalization on the industry, however, is the implications of digital technology on the renewable energy sector. Scaling renewable energy in the coming decades is necessary not only to help growing economies meet their energy requirements but also as a means to combat ongoing climate change. Renewable energy resources such as wind and solar, however, are highly variable and require highly accurate and timely forecasting to ensure grid stability.[31] Near real-time data are required for forecasting, and advanced, automated grid control is necessary for switching between energy sources during periods of low production. Digital technologies are becoming the enabling force allowing these requirements to become a reality.

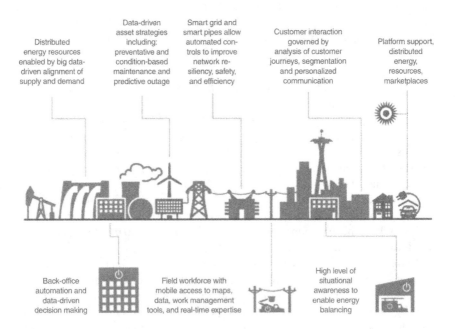

Figure 3.3 The Digital Energy Utility of the Future

Source: Adapted from Booth, Mohr, and Peters (2016).

Due to its intermittent and distributed nature, managing renewable energy resources requires expanding visibility, intelligence, and control (e.g., deploying connected, intelligent, and secure devices; automation equipment; sensors and smart meters). With digital technologies providing higher-resolution data sets, better algorithms, and new modeling tools, actionable intelligence can be provided to grid managers, mitigating the risks of intermittent energy production.[32] IoT technologies, AI, and machine learning can sense changing grid conditions (e.g., changes in demand or production) and quickly take appropriate actions (e.g., redirecting more produced energy to storage; shifting to a different, more available energy source).

While large-scale grids benefit in numerous ways, microgrids are using digital tools and IoT technologies to gain their share of the energy sector. As energy disruptors increase, such as natural disasters, increased demand on large-scale grids, and even IoT itself, so does the propensity for power outages and blackouts. Microgrids allow individuals, buildings, and even small communities to provide their own energy or function as a contributing source to a larger grid. Digital sensors and intelligent connections allow this to occur safely and reliably by automatically shutting off or reversing the flow of energy from a microgrid to the larger grid. Additionally, microgrids themselves can be a combination of renewable energy and traditional sources, such as solar power and diesel fuel, and switch from one to the other as needed through digital tools.[33] In Puerto Rico, where more than a third of the population was without power for months after Hurricane Maria in 2017 and with the entire island being rattled by earthquakes, microgrids and small, localized energy sources provide an important stabilizing factor into the larger grid.[34]

In addition to overall grid management, digital technologies are also revolutionizing the management and operation of renewable energy plants (e.g., wind and solar farms, hydropower production facilities). For example, at wind farms, kilowatt hours are lost every year due to unplanned downtime, operational inefficiencies, and inaccurate forecasting. Schedule-based maintenance is also largely inefficient and cuts years off the lifetime of turbines. IoT and other digital technologies, however, are providing the data, analytics tools, and AI systems to make turbines smarter and more productive. Combined sensor data and AI analytics make turbines more reliable and able to generate more energy. Data analytics programs can help operators react in real time to changing wind velocity and direction as well as grid demand, allowing for improved management of the intermittency of wind energy.

Brandon Owens, Digital Strategy and Innovation Director at GE Digital

Brandon Owens has worked at GE for over a decade and, in his recent years, has led the development of a digital transformation strategy throughout the GE enterprise. With the scope of GE reaching across sectors, Brandon notes that digitalization has been occurring at GE in healthcare, aviation, power, and renewable energy. To give an example, Brandon says,

> From GE Power's perspective, the digital wave is currently transforming the energy industry and it has impacted every facet of our business. There is enormous promise for digital technologies in energy, and also some pitfalls for organizations that can't figure out how to ride the [digital] wave. Digital technologies have the potential to accelerate deployment and adoption of both our generation and our transmission and distribution product offerings.

In another example, at GE Digital, the strategy has been to "develop Predix, an operating system for the Industrial Internet of Things (IIOT) built of Cloud Foundry open-source technology." Brandon elaborates on the technology by saying, "Predix provides features that allow third parties to develop and run applications. This has created an ecosystem of Predix developers and users that support GE's digital offerings."

When it came to overcoming barriers to digital adoption at GE, Brandon commented that

> Any transformational process is by nature disruptive and involves barriers. One of the primary barriers [GE] faced was determining how to help our customers digitize and what that meant for our product development and commercialization suite. Digital technologies have a vast potential, but there's a lot of white space right now and determining the right direction across a broad portfolio of businesses is challenging.

With vast digital offerings and opportunities, having a digital strategy is critical.

Lessons from the expert: The key to an organization's successful digital transformation is figuring out how to solve customers' toughest challenges with digital solutions, rather than simply encouraging customers to digitize for the sake of digitalization. Customers have real-world problems and face limited budgets – digital solutions must help customers succeed and make economic sense.

Sensors also play additional roles by detecting anomalies in turbine operation that may go unnoticed in regular inspections and alerting technicians to immediate issues, allowing maintenance to be performed on a case-by-case basis. Once identified, many malfunctions can even be addressed remotely, saving technicians from being sent to hazardous or remote locations. Digital twin technologies are providing another method for proactive maintenance by combining historical, physical, and real-time data to predict asset failure. Already deployed globally, digital wind farms are resulting in up to 10 percent reductions in maintenance costs and 3 percent increases in revenue.[35]

Digital technologies are transforming solar and hydropower production in much the same way as they are wind power. In solar and hydroproduction, digital twins simulate how a plant should operate, helping managers fine tune operations for optimal performance and flagging potential maintenance needs before asset failure occurs, avoiding unplanned downtime, and bypassing regular maintenance visits. Digital technologies are improving turbine/panel performance, reducing losses, and optimizing asset management across the renewable energy sector. With digitalization comes the potential to enable true "smart" grids using battery storage and the "smart" release of renewable energy, all while remaining responsive to end users' needs and synchronizing production with weather forecasts and flow control.

As digitalization continues to transform the renewable energy sector and smart grids become more of a reality, the potential of green energy is being unlocked to meet the growing demand for electricity due to population growth, circumnavigate the constraints of centralized power supplies to provide electricity to rural and underdeveloped communities, and combat the rising threat of climate change.[36] In this way, the experience of the renewable energy sector provides an example of how digital technologies not only provide for improved operations, efficiencies, and savings within an industry but also have the potential to benefit both society at large and the greater environment by providing resilience, security, and sustainability.

Notes

1 B. Kelechava (2016). *Water infrastructure problems worldwide.* American National Standards Institute. https://blog.ansi.org/2016/06/water-infrastructure-problems-worldwide/#gref
2 ICTpost (2015). *Technology leapfrog effect: Innovating the energy sector in India.* http://ictpost.com/technology-leapfrog-effect-innovating-the-energy-sector-in-india/
3 M. Orsero (2019). Technological leapfrogging and development: The example of Kenya. *The Perspective.* www.theperspective.se/technological-leapfrogging-and-development-the-example-of-kenya/
4 M. Jacobson (2016). The developing world can leapfrog dirty coal and go straight to clean energy. *Fast Company.* www.fastcompany.com/3056313/the-developing-world-can-leapfrog-dirty-coal-and-go-straight-to-clean-energy
5 J. Sperling (2019). *Urban-rural nexus science across scales for a 'coupling' and 'leapfrogging' of integrated services for smart-healthy-resilient cities, communities, to regional competitive advantage.* Iowa State University: SUS-RURI Proceedings. https://sus-ruri.pubpub.org/pub/sperling
6 World Economic Forum (2017). *Digital transformation initiative.* World Economic Forum.
7 N. Ismail (2019). *Digital transformation in the telecom industry: What's driving it?* Information Age. www.information-age.com/digital-transformation-in-the-telecom-industry-123478152/
8 R. Kalakota (2019). *Transform telecom: A data-driven strategy for digital transformation.* www.hitachivantara.com/en-us/pdfd/white-paper/digital-transformation-of-telecom-industry-liquid-hub-whitepaper.pdf

9   S. Arustamyan (2018). How the telecom sector is developing in times of digital transforma-
    tion. *Telecoms.* www.telecomstechnews.com/news/2018/mar/19/how-telecom-sector-
    developing-times-digital-transformation/

10 R. Kalakota (2019). *Transform telecom: A data-driven strategy for digital transformation.*
    Hitachi. www.hitachivantara.com/en-us/pdfd/white-paper/digital-transformation-of-
    telecom-industry-liquid-hub-whitepaper.pdf

11 J. Stoughton (2018, April). Turning data into insights: How digitization creates
    new opportunities for the telecommunications industry. *D!gitalist Magazine.* www.
    digitalistmag.com/iot/2018/04/23/digitization-creates-new-opportunities-for-
    telecommunications-industry-06093584/

12 R. Garcia, G. Lenz, S. Haveman, & G. M. Bonnema (2019). State of the art of mobility
    as a service (MaaS) ecosystems and architectures – An overview of, and a definition,
    ecosystem and system architecture for electric mobility as a service (eMaaS). *World Elec-
    tric Vehicle Journal*, 11(1), 7.

13 A. Cartenì (2020). The acceptability value of autonomous vehicles: A quantitative anal-
    ysis of the willingness to pay for shared autonomous vehicles (SAVs) mobility services.
    Transportation Research Interdisciplinary Perspectives, 8(100224). https://doi.org/10.1016/j.
    trip.2020.100224

14 D. Benton (2017). *Transforming transportation with real-time analytics.* Supply Chain. www.
    supplychaindigital.com/technology/transforming-transportation-real-time-analytics

15 B. Chen, A. Baur, M. Stepniak, & J. Wang (2019). *Finding the future of care provision: The
    role of smart hospitals.* McKinsey & Company. https://healthcare.mckinsey.com/finding-
    future-care-provision-role-smart-hospitals

16 The impact of digital transformation in healthcare (2021). *LeewayHertz.* www.
    leewayhertz.com/digital-transformation-in-healthcare/.

17 Ibid.

18 M. Reddy (2020). *Digital transformation in healthcare in 2020: 7 key trends.* Digital Authority
    Partners. www.digitalauthority.me/resources/state-of-digital-transformation-healthcare/

19 Ibid.

20 Alliance of Advanced Biomedical Engineering (2018). *Future of smart hospitals.* https://
    aabme.asme.org/posts/future-of-smart-hospitals

21 M. Reddy (2020). *Digital transformation in healthcare in 2020: 7 key trends.* Digital Authority
    Partners. www.digitalauthority.me/resources/state-of-digital-transformation-healthcare/

22 Alliance of Advanced Biomedical Engineering (2018). *Future of smart hospitals.* https://
    aabme.asme.org/posts/future-of-smart-hospitals

23 Ibid.

24 D. Dumortier (2017). *Digital transformation and the rise of smart hospitals.* Healthcare. www.
    healthcareglobal.com/technology/digital-transformation-and-rise-smart-hospitals

25 M. Reddy (2020). *Digital transformation in healthcare in 2020: 7 key trends.* Digital Authority
    Partners. www.digitalauthority.me/resources/state-of-digital-transformation-healthcare/

26 D. Dumortier (2017). *Digital transformation and the rise of smart hospitals.* Healthcare. www.
    healthcareglobal.com/technology/digital-transformation-and-rise-smart-hospitals

27 *The history of online education* (2017, November 29). Peterson's. Retrieved February 9,
    2020, from www.petersons.com/blog/the-history-of-online-education/

28 P. K. Nanda (2019). *Centre launches ₹9,000 Crore plan to digitize education delivery.* Mint.
    www.livemint.com/news/india/centre-launches-9-000-crore-plan-to-digitize-educa
    tion-delivery-1550683075713.html

29 J. Watson (2020). *Six strategic steps to digital learning success.* Digital Learning Collaborative.
    www.digitallearningcollab.com/blog/six-strategic-steps-to-digital-learning-success

30 A. Booth, N. Mohr, & P. Peters (2016). *The digital utility: New opportunities and challenges.*
    McKinsey & Company. www.mckinsey.com/industries/electric-power-and-natural-gas/
    our-insights/the-digital-utility-new-opportunities-and-challenges

31 M. Fayazfar (2019). *Digital transformation puts clean energy goals within reach*. Smart Energy International. www.smart-energy.com/industry-sectors/policy-regulation/digital-transfor mation-puts-clean-energy-goals-within-reach/

32 Ibid.

33 K. Anderson, N. DiOrio, B. Butt, D. Cutler, & A. Richards (2017). *Resilient renewable energy microgrids*. National Renewable Energy Laboratory. www.livemint.com/news/india/ centre-launches-9-000-crore-plan-to-digitize-education-delivery-1550683075713.html

34 R. Walton (2017). New $17.6B plan would rebuild Puerto Rico's grid with renewa-bles, DERs. *Utility Dive*. www.utilitydive.com/news/new-176b-plan-would-rebuild-puerto-ricos-grid-with-renewables-ders/512805/

35 GE Digital (2019). *The future of renewables is digital*. www.ge.com/digital/blog/future-renewables-digital

36 DNV GL (2019). *Digitalization and the future of the solar energy industry*. DNV GL.

4   The digital water ecosystem

James Moore, business strategist and the first to repurpose the term "ecosystem" with respect to business, observed that "In a business ecosystem, companies coevolve capabilities around a new innovation."[1] Business ecosystems have thus become an essential part of successful businesses through joint business planning strategies, collective action programs, or coinvesting in innovative technologies and business models. Building business ecosystems that include diverse stakeholders such as nongovernmental organizations is perhaps the only viable strategy to solve "wicked problems." Whereas businesses traditionally maintained only transactional relationships, mature business ecosystems can unlock new opportunities and innovative solutions (Figure 4.1).[2]

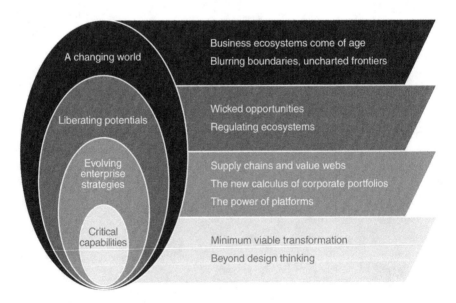

Figure 4.1 Business Ecosystems Come of Age

Source: Adapted from Kelly (2015).

DOI: 10.4324/9780429439278-5

4.1 What is a business ecosystem?

Just as in a natural ecosystem, public and private sector enterprises are finding themselves to be part of an increasingly interconnected web of market players. James Moore wrote in a 1993 Harvard Business Review article that "Successful businesses are those that evolve rapidly and effectively. Yet innovative businesses can't evolve in a vacuum. They must attract resources of all sorts, drawing in capital, partners, suppliers, and customers to create cooperative networks."[3] Indeed, similar to a natural ecosystem, a business ecosystem is a dynamic and coevolving community of diverse actors that create and capture new value through both collaboration and competition.[4] They create, scale, and serve markets in ways that a single organization or traditional industry could not achieve on its own. Whereas a business ecosystem can exist within a single company (e.g., Apple Inc. offers an "ecosystem" of products from laptops to smartphones to music devices), they are often a collection ("ecosystem") of firms operating in loose alliance.[5]

An illustration of a business ecosystem as a "living organism" is provided in Figure 4.2.[6]

Digital technology has been a major driving force connecting businesses and building them into an ecosystem of market players. Likewise, digital technologies are further advanced by the existing business ecosystems and strategic partnerships that enable creative thinking, piloting, and other collaborative initiatives promoting knowledge sharing and market penetration. Such collaboration enables public and private sector enterprises to develop innovative,

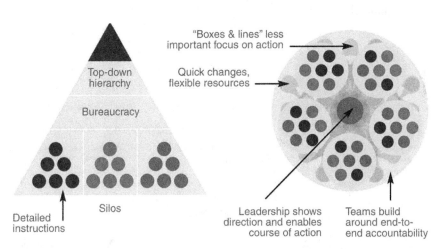

Figure 4.2 An Organization as an Agile Living Organism.

Source: Adapted from "How to leapfrog technology adoption cycles and gain rapid competitive edge" (n.d.).

cocreated solutions by utilizing each members' skills and resources.[7] They also accelerate learning and innovation and create new ways to address environmental and social challenges. Members of business ecosystems can be either large or small players, but they are all motivated by shared interests, goals, and values.

Through the development and expansion of business ecosystems, boundaries which previously defined the relationships, interactions, and possibilities of businesses are dissolving. The water sector especially is experiencing a breakdown of boundaries as organizations expand their web of relationships across industry lines in pursuit of solutions to the world's most pressing water challenges. As the following section shows, the digital water ecosystem includes diverse public and private sector participants, many of whom are developing unique partnerships, technologies, and digital approaches to address water scarcity, quality, and resiliency concerns, among others.

4.2 Solution providers

The digital water ecosystem is expanding as the sector's vast market players turn to mutually beneficial partnerships in their search for innovative, digital solutions. Utilities and industrial players are increasingly partnering with emerging technology providers to address operational efficiency and conservation concerns. Likewise, the agricultural sector is seeing a rise in the use of digital technologies both as a part of corporate supply chains and with individual farmers. Meanwhile, technology hubs and accelerators, as well as strategic investors and academic institutes, are providing the mentorship, funding, and research to connect market players, develop new technologies, and launch solutions. Together, these public and private sector enterprises comprise the digital water ecosystem and are the foundation of solutions enabling better water management (Figure 4.3).

WATERSHED/SUPPLY CHAIN | **INDUSTRIES, WATER AND WASTEWATER UTILITIES** | **CUSTOMERS/CONSUMERS**

- Real-time water quantity and quality monitoring
- Predictive analytics
- Integrated water, energy and agriculture management

- Predictive analytics
- Water quality monitoring
- VR/AR augmented workforce
- Smart hardware
- AI managed assets

- Customer engagement and conservation analytics
- Demand forecasting
- Smart home and hydration

Figure 4.3 The Digital Water Value Chain

Source: Water Foundry Advisors, LLC (2019).

Utilities

There are larger movements in the realm of digital asset solutions for water utilities now than ever before. Over the past year plus, we have seen the utility sector increasing scale digital water technologies in response to COVID-19. The "digital utility" of the future is upon us and illustrated in Figure 4.4.

For example, during 2019, the company Innovyze, based in Portland, acquired Emagin in a promising development for increased AI applications in the water sector. Innovyze specializes in water infrastructure data analytics, while Emagin combines AI and machine learning software to help utilities better manage their daily operations. The two companies share the same vision to help revolutionize water utilities through groundbreaking software. Their combined effort nearly doubles the size of Emagin's current team in the United Kingdom and bolsters efforts to legitimize and spread this technology further.[8]

On the consumer side, software continues to become more sophisticated and less burdensome for the typical user. One example is the company WaterSmart Software, acquired by VertexOne in mid-2020, a leading provider of enhanced SaaS solutions to the utility industry. WaterSmart whose software offers self-service solutions for leaks, payments, and other customer concerns, providing relief for water utilities and allowing consumers that much

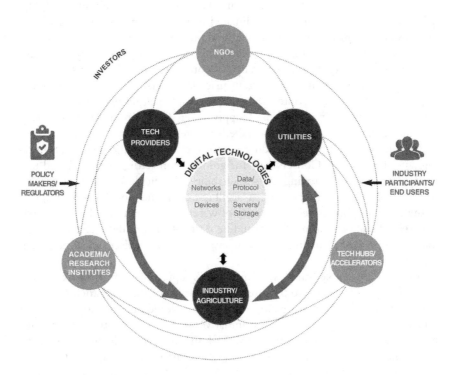

Figure 4.4 The Digital Water Utility Ecosystem

Source: Adapted from Sarni *et al.* (2019).

more control.[9] The incentive for a utility to adopt such a technology is not only the desire for a better experience for their customers but also the ability to reduce operating costs and capital expenses via digital engagement. As of April 2020, WaterSmart has strategically partnered with Master Meter to offer more potent actionable insights to utility operators. Master Meter is already known and established for their smart network and analytics, specializing in meter data management software, which are now complemented by Water-Smart's customer engagement platform. Strategic partnerships like these are becoming more commonplace in the market and are a smart tactic for promoting multiple technologies while strengthening each company's standing.[10]

More on acquisitions in the water sector – utilities and industrial applications later in this chapter.

Agriculture

All food requires water, and roughly 70 percent of annual freshwater usage is for food production alone. Thus, innovations in water management cannot ignore the agricultural sector as part of the digital water ecosystem. In terms of digital water solutions, companies with large agricultural components of their supply chains may be some of the first targets for creating large-scale impacts across food supply chains. As companies look to increase their sustainability, decrease business risk, and address climate impacts, many are already turning to digital water solutions for their agricultural providers.

One example of a company leading the way in agricultural supply chain sustainability is Cargill, who has recently committed to achieving sustainable water management in all their current priority watersheds across their supply chains by 2030 using a context-based approach.[11] Cargill plans to undertake this by assessing their agricultural supply chain for exposure to water scarcity and maintaining priority watersheds where they operate. They have also engaged in partnerships with leading NGOs such as the World Resources Institute and The Nature Conservancy for guidance on setting science-based targets[12] and support improving the sustainability of agriculture in key countries, respectively.[13] With this approach, Cargill considers the global water footprint of the company alongside the severity of water challenges in the local context. In terms of scale and impact, it would be difficult to consider a change of this magnitude – yet maintaining local relevance – taking place without digital solutions.

An applied example can be seen in Cargill's supply chain in Nebraska where beef producers use water to irrigate the row crops for their livestock feed.[14] In cooperation with The Nature Conservancy and Nestle Purina, Cargill targeted this water use for a water reduction initiative. In May 2018, a three-year water project was launched to improve the water footprint of Cargill's beef producers. The project employed the use of cutting-edge smart weather sensors to drastically reduce the amount of water used in irrigation and had an estimated offset of 2.4 billion gallons of irrigation water saved over three years.

Anheuser-Busch InBev (AB InBev), a global leader in both brewing and water stewardship, has also turned to digital technologies for agricultural applications along its supply chain. In 2019, AB InBev partnered with Sentera, an agronomic insight and technology provider, to improve productivity and reduce environmental impacts among their barley growers. By partnering with Sentera, AB InBev utilizes real-time and historic data, including satellite imagery, drones, and other smart devices alongside weather and soil data, to build crop models and improve agricultural practices.[15] Again in 2020, AB InBev deployed a new digital supply chain solution, this time by partnering with SettleMint, a blockchain-as-a-service platform.[16] The blockchain technology, where piloted, will give full transparency and traceability in AB InBev's barley supply chain. This transparency will ensure the quality and integrity of raw ingredients as well as the efficient use of natural resources, including water.

Both Cargill and AB InBev provide an example of industry leaders in agriculture working to address "wicked water problems" to solve the complex puzzle of water management along the agricultural supply chain worldwide. In these cases, the food and beverage sector giants act as top-down drivers of digital adoption, with the corporate need for increased yields, efficiency, and accountability inspiring innovative projects along their supply chains. However, the growing demand from independent farmers for water-saving solutions is opening the market now to companies which can provide unique technology solutions tailored to the needs of the agricultural industry. Several examples of emerging companies aiming to secure a spot in this space include Arable, CropX, and other precision agriculture companies.

Irrigation is one of the most problematic issues to solve in terms of agricultural water use worldwide. Crops require immense volumes of water, but at the same time, inefficient and conventional irrigation systems often contribute to large amounts of wasted water from imprecise irrigation and overwatering. Arable is one of the leading companies in precision agriculture exploring several large avenues where technology can assist in informed irrigation and agricultural decision-making.[17] Those areas include irrigation, weather, digital insights, research and development, and API integration. Arable's paired web/app interface displays data from field sensors and local weather stations on water stress, crop evapotranspiration, soil and plant conditions, precipitation, and weather forecasts to inform irrigation schedules, monitor crop growth, and predict harvest times.

Another competitor in the market is CropX, which also seeks to add value by saving farmers time and resources via their actionable insights.[18] The CropX technology boasts over 30 percent demonstrated water savings in the field. This technology uses a cloud platform connected to an app to provide data from several sources all at once. CropX offers soil sensors for collecting moisture, temperature, and electrical conductivity data at multiple depths. Likewise, weather information is gathered from several agriculture-specific weather data sources that are combined to form CropX's algorithms. Ultimately, CropX presents users with one-week forecasts for improved decision-making.

Additionally, CropX combines aerial imagery, topography maps, soil mapping, hydraulic models, crop models, and user input for the most comprehensive insights, alerts, and information available. CropX's standout feature is capitalizing and integrating soil data, which had not been done before with other precision agriculture tools. Both CropX and Arable were ranked in the 2020 THRIVE TOP 50 AgTech companies in the category of IoT and Software.[19]

In line with the aforementioned tools, it is worth noting that CropIn not only provides many similar features for farm management but also includes an element which is critical for farmers: managing their risk profiles throughout the season.[20] CropIn's predictive and prescriptive solution incorporates AI and machine learning, which are becoming more popular in every predictive and risk management tool. Overcoming risk is one of the farmers' largest challenges and, when ill-prepared for, can lead to the largest losses throughout the year. Understanding past conditions in order to predict future scenarios is therefore hugely important for farmers to make better decisions at the beginning of the year.

Industrial

Industrial users share many of the same challenges as large agricultural users in that they have huge global supply chains which often impact local water sources in a negative way. For this reason, pressure has been applied for private entities to step up and take responsibility for water management along their entire global supply chains. Nestlé, for example, has committed to pursue water stewardship by following the guidance and structure of a third-party standard.

Specifically, Nestlé has agreed to follow the Alliance for Water Stewardship's (AWS) program centered around the AWS Standard. According to the AWS, water stewardship can be defined as:

> *The use of freshwater that is socially equitable, environmentally sustainable, and economically beneficial, achieved through a stakeholder-inclusive process that involves site- and catchment-based actions.*

Within Nestlé, leaders plan to enact this vision via a roadmap toward responsible water management by incorporating AWS certification at various points across their global supply chain.

An example of their vision in action is observed at Nestlé's bottling facilities in California, all five of which are currently AWS certified.[21] By using advanced monitoring techniques and optimization of their overall water usage (specific technologies not disclosed), they were able to avoid using an estimated 204 million liters of water between 2016 and 2017 alone. Following this offset, their Cabazon facility in California was the first site in North America to be awarded gold-level certification for their positive contribution to the local groundwater system and high water quality standards. This provides a successful example where a large corporation with a significant supply chain impacts water by starting to move away from mere self-declarations of improvement and more toward achieving third-party standards.

For corporations like Nestlé, meeting the AWS Standard 2.0 at plants and facilities along their supply chain can be very difficult without implementing certain technologies. Likewise, monitoring at a high level is an absolute requirement for staying competitive against the Standard and against industry peers.

In another example, The Coca-Cola Company (Coca-Cola) has taken on an internal goal of replenishing the water supplies that they use in a timely manner.[22] They set about addressing that goal via partnerships with local NGO stakeholders that both assist in directly developing projects to replenish Coca-Cola's water supply and educating the company on what types of projects are valid toward their water stewardship goals.

The technological impetus from Coca-Cola is primarily focused on their need for massive wastewater treatment at scale to replenish their water supplies.[23] This has driven them to set their own quality standards at a high level for the water that they are returning to the environment. The high standards have led to a $1 billion investment across their entire system for the adoption of new processes and technologies that assist in meeting these ambitious wastewater treatment goals.

This is not dissimilar from Nestlé's efforts to follow the AWS Standard and with the difference between a high internal standard and a third-party standard becoming smaller and smaller; the question may be which becomes more valuable over time, and chances are that the third-party standards will triumph as they are more easily regulatable.[24] While the AWS Standard does not go into specific technology recommendations, it does provide a roadmap for each business process (e.g., understanding water challenges, committing to a water stewardship plan, implementation, evaluation, communication/disclosure) and water outcomes (e.g., water governance, water balance, water quality, and WASH) that can be improved. Technology providers will understand their niche and begin to target multiple areas of improvement at once with these large corporations in an overarching digital transformation strategy.

Notable solution providers are already rising in the industrial sector such as Aquassay, which in 2019 signed a major partnership with Ponticelli Frères to propose solutions to their clients to help with independence and performance under climatic variations.[25] Likewise, Nestlé began deploying Aquassay's real-time data SAAS solutions in 2019 across more than 20 countries.[26] The technology provides real-time data and trend analyses using predictive models to help bottling plants consume less water and produce high-quality water by monitoring every aspect of processes in their plants. The online application can visualize data on a smartphone, tablet, and computer to aid in decision-making such as troubleshooting and maintenance planning. This is just one of the technologies that Nestlé is employing to meet the AWS Standard and their water stewardship goals by 2025.

There are some surprising industrial players besides the usual suspects in water management. One is AT&T, for whom water plays a large role in their overall operations.[27] Water supply chain challenges have an impact on their business, despite their not producing water or food products. AT&T uses key performance indicators to measure their water usage per year. In terms of water conservation at their sites, they utilize the company HydroPoint, which provides smart water management solutions to remotely monitor and manage water systems in real time.[28] HydroPoint offers 360° Smart Water Management, a program that combines smart landscape irrigation, leak and flow monitoring, and professional services such as water bill analytics and 24/7 monitoring to assist in running systems at the industrial scale.

Tim Fleming, Director of Global Enterprise Sustainability at AT&T

Tim Fleming has held a range of positions in his twenty years at AT&T but in his current role, Tim leads initiatives at the intersection of technology and sustainability to identify digital solutions for reducing the environmental footprint of both AT&T and their customers. Projects have included utilizing smart water meters to enhance water use in cooling towers; installing HydroPoint smart irrigation controllers to reduce water use at facilities; and deploying Enterprise Building Management Systems to monitor equipment, minimizing mechanical cooling needs and water consumption. Tim also highlights the role of the AT&T network in facilitating water conservation through IoT connectivity for monitoring water utility infrastructure and enabling smart irrigation systems, among others. "At their core," Tim says, "these projects were all about saving water, reducing costs, and increasing efficiency across our systems. As technology advances, with 5G for example, we will continue to see more use cases to drive efficiency and reduce natural resource use."

Tim emphasizes the need to build an ecosystem of digital partners and encourages peers to identify and connect with businesses in other sectors that might have similar challenges regarding their primary water use drivers. He notes that

> even competitors can collaborate through member organizations. For example, we are members of the Global Enabling Sustainability Initiative (GeSI) where we have supported initiatives to identify how digital systems can drive positive outcomes across the UN Sustainable Development Goals including those related to water.

Lessons from the expert: For those just beginning their digital journey, Tim recommends starting small and keeping your baseline in mind to measure progress against your digital goals.

Technology hubs/accelerators

When dealing with emerging technologies, one cannot ignore the source of fertile creativity, which is often the entrepreneurs pushing start-ups into the field and the technology hubs and accelerators that support their growth. One such initiative is the Techstars Sustainability Accelerator created in partnership with The Nature Conservancy.[29] Techstars is based in Denver, Colorado, and runs annually. The goal of the program is to support purpose-driven, for-profit entrepreneurs who want to address key sustainability challenges.

From the 2019 Techstars cohort, the start-ups that emerged as most promising in water technology included 2NDNATURE, a company that has developed the first enterprise cloud stormwater management and compliance platform to help cities and counties effectively comply with the Clean Water Act.[30] This tool is also able to help prevent the further degradation of water quality in urban waterways. Another company, AQUAOSO Technologies, offers advanced water risk analytics to help financial institutions, insurers, investors, corporations, and policymakers manage their water risk profiles through a software platform and access to independent research and report development services.[31] This is also a useful tool for the agricultural sector, and it is being implemented across the Western United States to help growers understand water risk and assess water assets. Currently, the company claims 80 billion dollars' worth of agricultural assets monitored as well as 2.03 million acres researched and over 4,400 reports completed for clients to date.

Mammoth Trading is another Techstars participant with a software solution for the agricultural sector. Their platform, TAPP H20, allows producers to easily track and trade their water allocations, submit regulatory compliance documentation, and quantify their conservation practices.[32] TAPP H20 is a phone app that helps users easily access regulatory reporting and performance insights, trade water in a marketplace, and utilize water management training. Targeted users are growers, water managers, and corporations, all of whom can save time and money with Mammoth Trading's solutions. Thus far, there have been 500 subscriptions to TAPP H20 covering 60,000 irrigated acres with an additional 524,000 irrigated acres covered in their advanced surface and groundwater market service.

Finally, a third Techstars start-up addressing water, Gybe is a company that uses satellite imagery and proprietary ground-based hardware to transform the management, conservation, and restoration of aquatic ecosystems.[33] Gybe makes it easier to quantify water pollution, determine pollution sources, and continuously monitor critical watersheds. Some of their current and past users include AB InBev, the city of Salem, and The Nature Conservancy, where they are helping to develop new tools to accelerate watershed restoration projects. They are currently active in the Mississippi River Delta in Louisiana and are also developing a new project in North Carolina, while exploring similar opportunities across the United States.

Another leader in the accelerator and tech hub space is Imagine H20, a program that has been promoting innovation specific to the water sector since

2009.[34] Imagine H20 has supported over 100 start-ups with scale-up resources, and over 1,500 entrepreneurs from more than 40 countries have applied to be part of their competitive program. Thus far, over $500 million has been raised to support the participating start-ups. The value of the Imagine H20 program lies in their unique mentorship and participation in global water events such as the high-profile World Water Week in Stockholm. Another innovative aspect of their program is their Urban Water Challenge, which aims to improve the way urban water is sourced, cleaned, distributed, used, treated, and reused.[35] The 2019 Urban Water Challenge finalists included Indra, a company offering modular, decentralized wastewater treatment systems based in Mumbai, India; Stormsensor, a cloud connected sensor network for stormwater management – essentially a Google map for sewers – based in Seattle; and Smarter Homes, a company offering smart water meters and automated leak prevention system for high-rise apartments.[36] Smarter Homes, also headquartered in Mumbai, India, recently partnered with Namibia Water Corporation Ltd. To deploy 3,000 battery-powered smart water meters with integrated leak prevention systems that can reduce household consumption by approximately 35 percent.

These types of accelerators are critical for breeding and nurturing talent and ensuring that start-ups are competitive when they go to market. Likewise, accelerators and hubs can connect technologies with clients more quickly than through normal market entry routes. Established hubs/accelerators, such as Imagine H20, have large supporters and funders who can facilitate rapid growth, bringing a technology quickly to its next maturity stage.

Investors and digital water technology acquisitions

No brilliant technology can truly succeed without some form of funding at every stage of development. Fortunately, there are players in the funding field who are interested in making ideas a reality and driving the water sector forward. The digital transformation will not be cheap, especially for the water sector and those sectors which have large and costly infrastructure to update and maintain. However, the prospective savings ensure that the upfront costs are always worth it.

One such trusted financial player is XPV Water Partners, which invests in companies that positively impact water resources and their related processes both directly and indirectly.[37] XPV has increasingly focused on investing in digital technology companies. For example, XPV invested in SmartCover that offers a solution via proactive, remote monitoring for detecting and preventing sewer spills.[38] This technology has many variations for monitoring rainfall, vault water infiltration, intrusion detection, and tidal data, among other parameters.

Another XPV digital technology company was Aquatic Informatics which in 2020 was acquired by Danaher Corporation. Aquatic Informatics software supports fully remote operations for cities, agencies, and industries which need to monitor and manage water systems from home.[39] Existing customers had the opportunity to upgrade their existing software to cloud-based versions with a higher level of user control over data access and security. Their platform allowed users to maintain a high level of productivity during emergency times through enhanced data analytics and event notifications, as well as automated data collection and organization features.

These are several examples of technologies that are being supported by large funds for impact. When a start-up enters a partnership with an entity such as XPV, they go through the creation of a new strategic plan to accelerate growth. They use their expertise to try and differentiate the product on the market and generate increased demand. They usually go further than this as well, by helping the company to connect with new customers, hire new talent, pursue mergers and acquisitions, and improve their margins. XPV, for example, helped Aquatic Informatics conduct calculated growth and expansion. XPV relied on their resources including extensive databases, industry connections, and sector expertise to help identify the most promising deals for Aquatic Informatics going forward. Step-by-step Aquatic Informatics is on its way to becoming the world's largest water data management company.

Another recent development in the value placed on digital water technology companies was the acquisition of Innovyze by Autodesk.[40] On March 1, 2021, Autodesk, Inc. announced it signed a definitive agreement to acquire Portland, Oregon-based Innovyze, Inc., a provider of water infrastructure software, for $1 billion net of cash subject to working capital and tax closing adjustments. According to the press release, "the acquisition positions Autodesk as a technology leader in end-to-end water infrastructure solutions from design to operations, accelerates Autodesk's digital twin strategy, and creates a clearer path to a more sustainable and digitized water industry."

> Innovyze's modeling, simulation, and predictive analyses solutions enable more cost-effective and sustainably designed water distribution networks, water collection systems, water and wastewater treatment plants, and flood protection systems. Further, Innovyze's solutions centralized infrastructure asset visibility to optimize capital and operational expenses. Combining Innovyze's portfolio with the power of Autodesk's design and analysis solutions, including Autodesk Civil 3D, Autodesk InfraWorks, and the Autodesk Construction Cloud, offers civil engineers, water utility companies and water experts the ability to better respond to issues and to improve planning.

Another significant player in the water investment is Water Asset Management, which aims to be a recognized leader in managing global investments that solve water quality and availability issues.[41] Their team is composed of experienced water professionals who can identify investable trends across the global sector. They invest not only in emerging technologies but also in public and private equity in regulated utilities, water resources, infrastructure, treatment, and technology solutions.

Some of Water Asset Management's investment vehicles include the TRF Master Fund, the WAM Global Water Equity, and the Water Property Investor.

The latter, for example, invests in a diverse portfolio of water-rich agricultural properties with the potential to develop long-term sustainable water supplies. For strictly private equity investments, Water Asset Management uses the following vehicles: Water Property Investors, US Water and Land, and Southwest Water Company. These types of large investment bodies can secure water sources and procure the funds necessary for the infrastructure overhauls that are soon to be needed all over the country. Specifically, Water Asset Management is concerned with the growing water management issues in the western region of the United States where scarcity is driving these different investing arms to take proactive measures against risk to try and secure future returns. The strategic advantage of diversifying long-term water supplies is to increase the sustainability and resilience of water supplies.

Also, on World Water Day in 2021, the first place-based water technology investment fund focused on digital water technologies was launched.[42] Colorado River Basin Fund goals are to advance innovative approaches to water challenges and engage a range of stakeholders, all who have an important role to play in solving them. The Colorado River Basin Fund will aim to meet these challenges and opportunities by identifying, investing in, and scaling early-stage water sector technology innovations in categories such as satellite data acquisition and analytics, IoT and edge computing, artificial intelligence and water smart agriculture, homes, and cities. We view the Colorado River Basin as a strategic testbed of sorts to determine the feasibility of emerging technological solutions for subsequent application in the global water sector providing an opportunity for scale.

Academia

Universities and research institutes also play a major role in the exploration and development of digital technologies and other innovations for the water sector. With access to unique funding sources, inquisitive minds, and widespread industry connections, universities and research institutes are distinctly placed to accelerate the adoption and development of digital water technology. Based upon a 2013 report, some of the highest ranked universities for water technology research include the National University of Singapore, Nanyang Technological University, Delft University, the University of California Davis, Wageningen University, and the Swiss Federal Institute of Technology.[43]

Singapore and the Netherlands tend to lead the way consistently for water innovation, especially in the areas of desalination, reuse, drinking water and wastewater, and nutrient recovery. The National University of Singapore has outstanding research in water and wastewater treatment technology, membrane technology, and water resource management. In water and wastewater treatment technology, they conduct research on pilot-scale membrane bioreactors and integrated microbial electrochemical sensor systems, which are developed to continuously monitor toxic compounds present in the water that could impact water treatment processes in PUB (Singapore's water agency) water reclamation plants.

Another top university in the list is Delft University in the Netherlands, featuring the prominent IHE Delft Institute for Water Education, which is the largest graduate water education facility in the world.[44] Their research relates to safe drinking water and sanitation, water-related hazards and climate change, water and ecosystems quality, water management and governance, water food and energy security, and information and knowledge systems. In addition to generating research and technology, they also focus on the larger picture in terms of capacity development and institutional strengthening. As is well known by professionals in the water technology field, these factors are equally as important as the technology itself for success.

The University of Exeter is another leading university in terms of academic institutions which are investing in developing emerging water technologies. Their Centre for Water Systems produces research and development in the following areas: water distribution system management; urban drainage system management; water, climate change, and sustainability; artificial intelligence research and applications; water resources management; and flood risk management. An example project includes iWIDGET, which was successfully completed in 2015 as a European Commission project.[45] It was created with the use of ICT technologies for integrated supply–demand side management. This project was led by Professor Dragan Savic, a founder of the Centre for Water Systems at the University of Exeter. Dragan is also currently the CEO of the KWR Water Research Institute, a nonuniversity affiliated institute also in the Netherlands. KWR performs its own research on digital technologies but has also developed a research initiative that is cofunded by water utilities in the Netherlands. Through the program, KWR connects technology providers to water utilities which can then test new or implement proven technologies. By collaborating with market players across the water sector, KWR facilitates cutting-edge research and the commercialization of game-changing digital technologies.

NGOs

While there are numerous small- to medium-sized NGOs that deal with water-related issues across the United States and globe, not all deal directly with technical issues and driving technological advancement forward. One organization that is focused on such issues is the Water Research Foundation (WRF), one of the leading research organizations to advance water science.[46] WRF funds, manages, and publishes research on the technology, operation, and management of drinking water, wastewater, reuse, and stormwater systems. Their primary topics include biosolids, climate change, cyanobacteria and cyanotoxins, disinfection by-products, energy optimization, intelligent water systems, microbes and pathogens, and resource recovery.

Some featured WRF technologies include Lystek Thermal Hydrolysis, which is a cost-effective biosolids treatment solution for helping water utilities produce more biogas and reduce their biosolids volumes, costs, odors, and greenhouse gases.[47] The system is usually installed as a postanaerobic digestion and post-dewatering solution. It is a scalable, flexible solution and can be deployed either for an individual wastewater treatment plant or installed regionally to serve multiple customers at once. Another featured technology is the AirPrex sludge optimization process from CNP-Technology Water and Biosolids Corp., which removes and recovers orthophosphate via the controlled formation of struvite.[48] The benefits of this particular technology include over 90 percent orthophosphate removal from sludge, mitigation of uncontrolled struvite precipitation in pipes and process equipment, improved dewaterability of biosolids, reduction of polymer consumption by sludge dewatering processes, and the harvesting of struvite for use as a slow-releasing phosphorus fertilizer.

In addition to internally supporting new technologies, the WRF also offers the Leaders Innovation Forum for Technology, which is a multipronged initiative undertaken by the WRF and the Water Environment Federation to assist in bringing new water technology to market more quickly and efficiently. This involves evaluating the technology via demonstrations, boosting communications to include training, education, and outreach designed to promote innovation, benchmarking utilities based on identification of resources and policies needed, and an informal forum for the research and development phase.

In terms of NGOs, you cannot ignore organizations such as the Nature Conservancy (TNC) and World Wildlife Fund (WWF) for their large size and scope of work in water. As already mentioned, TNC plays a role in the digital water ecosystem both by directly partnering with communities/farmers and in its position as a cocreator of the Techstars Sustainability Accelerator through which it has partnered with several digital water companies (e.g., Gybe and Mammoth).

Likewise, the WWF has also begun exploring digital solutions in recent years. WWF and partner organizations have introduced revolutionary blockchain technology for tracking fish and eliminating illegal fishing.[49] In addition, drone photography and digital tags have been used to study whales for other WWF conservation initiatives.[50] Beyond their potential for tracking and quantifying wildlife, WWF has recognized the potential for digital technologies to safeguard watersheds and address water scarcity.[51] For example, after the WWF-Canada 2017 Watershed Reports identified high threats to urban watersheds in Canada and large gaps in water data, the WWF sponsored the Generation Water Tech Challenge. At the challenge's conclusion, four award recipients were chosen with solutions focusing on data management, sensor technology, and AI.[52]

Public sector

The public sector is active in advancing new water technologies and research alongside the private sector, and oftentimes, they work in tandem. While there are outstanding private sector initiatives – from utility/technology provider partnerships to cutting-edge research at water institutes, there are also public sector initiatives that can assist in advancing water science. One of those initiatives in the United States is the California Open Data Initiative. The California Open Data Initiative promotes unlocking government data to the public for help in supporting research and innovation as well as improved water management.[53] The initiative consolidates and standardizes data from numerous state agencies across the state. These data, now more complete and easily accessible, can then be used by other public or private entities to inform decision-making, planning, and innovation. In addition to water, the California Open Data Portal also offers usable data in the categories of: natural resources, health and human services, economy and demographics, government, transportation, and now COVID-19.

While the aforementioned initiative deals with many topics besides water, Digital Water by the UK Water Partnership is focused specifically on driving digital innovation within the UK water sector.[54] The United Kingdom also has many forward-looking water utilities, so it is no surprise that this initiative was naturally formed to move the entire nation forward. The partnership brings together diverse water sector organizations to promote understanding, cooperation, and coordination. They have recently published a White Paper and Call to Action which calls for more focused research on promising commercial opportunities for digital water technologies in the United Kingdom. This includes recommendations for increased collaboration across the UK water economy and more active marketing of UK digital water expertise worldwide.

Digital water innovations are also being taken very seriously across the European mainland. The European Union sponsors many initiatives that deal with the environment and technological innovation. One of them is the ICT4 Water Cluster, which is a hub for EU-funded research projects on ICT and water management.[55] Several of their active projects include the AfriAlliance, a project which aims to better prepare the African continent for climate change by supporting networks in identifying technological solutions for key water challenges.[56] Another project is Aqua3S, which combines technologies in water safety and security toward standardization of existing sensor technologies that can be complemented with state-of-the-art detection mechanisms.[57] AquaNES is a different project that offers an improved combination of natural and engineered components, providing an example of the wastewater sector's innovative trend toward hybrid gray-green treatment techniques.[58] They currently have thirteen demonstration sites in Europe, India, and Israel that represent a wide diversity of regional, climatic, and hydrogeological conditions. The sites include technologies such as bank filtration, managed aquifer recharge, and constructed wetlands, plus engineered pretreatment and posttreatment options.

In addition to these types of clustered projects, there are also many innovative conferences around the world that are good meeting grounds for the public and private sectors to intersect and swap ideas as well as make valuable business connections. This year due to the COVID-19 pandemic, many of these conferences have been cancelled or postponed, but some saw the opportunity to turn themselves into a fully digital experience. One of those conferences is Water Innovation Europe. They are shifting their conference into an online version that includes plenary sessions, digital working group meetings, digital networking, and digital exhibition opportunities. The conference includes water innovation awards to raise the visibility of innovation approaches and technological solutions with high market potential, a feature typical of all water conferences.[59] Winners from 2020 include Idrica, Deltares, bluephage, De Montfort University Leicester, and WeCo.

On the highest level of the public sector in Europe is Water Europe (WE) run by the EU. This is an overarching group for technology and innovation with 216 members and counting from the global water sector.[60] WE uses a combination of ambassadors, working groups, and events to coordinate their program for change. An innovative feature of WE are their Water Vision Leadership Teams, which are deployed to bring successful innovations to market in Europe and beyond. These teams independently assess water challenges and solutions ranging from digital water solutions, addressing the true cost of water, and water treatment technologies to hybrid and green infrastructure, water-smart industries, and water solutions for both city and rural applications. WE's involvement in other programs such as the Water-Oriented Living Labs, the European Junior Water Programme, and the International Water Dialogues provides additional platforms for developing and testing combinations of solutions in the field, advancing young water professionals, and arranges business and research collaboration between the EU and other strategic global regions, respectively.

Notes

1 J. F. Moore (1993). Predators and prey: A new ecology of competition. *Harvard Business Review.* https://hbr.org/1993/05/predators-and-prey-a-new-ecology-of-competition
2 E. Kelly (2015). Business ecosystems come of age. *Deloitte Business Trends Series*, 49(2), 28–35. https://www2.deloitte.com/content/dam/insights/us/articles/platform-strategy-new-level-business-trends/DUP_1048-Business-ecosystems-come-of-age_MASTER_FINAL.pdf
3 J. F. Moore (1993). Predators and prey: A new ecology of competition. *Harvard Business Review.* https://hbr.org/1993/05/predators-and-prey-a-new-ecology-of-competition
4 E. Kelly (2015). *Business ecosystems come of age.* Deloitte Business Trends Series. https://www2.deloitte.com/content/dam/insights/us/articles/platform-strategy-new-level-business-trends/DUP_1048-Business-ecosystems-come-of-age_MASTER_FINAL.pdf
5 Ibid.
6 How to leapfrog technology adoption cycles and gain rapid competitive edge (n.d.). https://cdn2.hubspot.net/hubfs/5078049/Media_files/Whitepapers_How_to_leapfrog_technology_adoption_cycles.pdf
7 E. Kelly (2015). *Business ecosystems come of age.* Deloitte Business Trends Series. https://www2.deloitte.com/content/dam/insights/us/articles/platform-strategy-new-level-business-trends/DUP_1048-Business-ecosystems-come-of-age_MASTER_FINAL.pdf
8 M. Simpson (2019). Waterloo's Emagin acquired by Innovyze to improve water infrastructure data. *Betakit.* https://betakit.com/waterloos-emagin-acquired-by-innovyze-to-improve-water-infrastructure-data/
9 WaterSmart (2020). www.watersmart.com/
10 Water Finance & Management (2020, April). *Master meter, WaterSmart announce collaboration.* https://waterfm.com/master-meter-watersmart-announce-collaboration/
11 Cargill (2020). *Cargill commits to restoring 600 billion liters of water by 2030.* www.cargill.com/2020/cargill-commits-to-restoring-600-billion-liters-of-water-by-2030
12 Ibid.
13 Cargill (2018). *Cargill and the nature conservancy.* www.cargill.com/doc/1432101097009/cargill-tnc-partnership-fact-sheet-pdf.pdf

14 Cargill (2019). *4 ways Cargill is saving water in our operations, supply chains and communities.* www.cargill.com/story/world-water-week-four-ways-cargills-pledge-to-saving-water

15 W. Vogt (2020). *Bringing high tech to barley production.* Farm Progress. www.farmprogress. com/crops/bringing-high-tech-barley-production

16 ABInBev (2020). *From barley to bar: AB InBev trials blockchain with farmers to bring supply chain transparency all the way to beer drinkers.* https://ab-inbev.eu/news/from-barley-to-bar-ab-inbev-trials-blockchain-with-farmers-to-bring-supply-chain-transparency-all-the-way-to-beer-drinkers/

17 Arable (2020). www.arable.com/

18 CropX (2020). www.cropx.com/

19 THRIVE (2020). *SVG Ventures releases 2020 THRIVE TOP 50 AgTech and TOP 50 FoodTech reports.* https://thriveagrifood.com/svg-ventures-releases-2020-thrive-top-50-agtech-and-top-50-foodtech-reports/

20 Cropin (2020). www.cropin.com/

21 C. Galli (2018). *Water stewardship.* Nestlé. www.nestle.com/stories/nestle-aws-standard-water-stewardship

22 J. Nastu (2019). *Coca-Cola shares water management strategies: Q&A with Jon Radtke.* Environmental + Energy Leader. www.environmentalleader.com/2019/02/coca-cola-shares-water-management-strategies-qa-with-jon-radtke/

23 The Coca Cola Company (2020). *Wastewater: Safely returning the water we use to make our beverages.* www.coca-colacompany.com/sustainable-business/water-stewardship/treating-and-recycling-wastewater

24 C. Galli (2018). *Water stewardship.* Nestlé. www.nestle.com/stories/nestle-aws-standard-water-stewardship

25 Ponticelli (2019). *Partnership for optimising water management.* www.ponticelli.com/en/aquassay-2/#:~:text=Thierry Le Gangneux%2C the Ponticelli, contract to formalize their partnership

26 Aquassay (2019). *Partnership Nestlé waters / Aquassay.* https://aquassay.com/partnership-nestle-waters-aquassay/?lang=en

27 AT&T (2020). *Water management.* https://about.att.com/csr/home/reporting/issue-brief/water-management.html

28 HydroPoint (2020). *360° Smart water management technology.* www.hydropoint.com/

29 Techstars (2020). *Techstars sustainability accelerator in partnership with the nature conservancy.* www.techstars.com/accelerators/sustainability

30 2NDNATURE (2020). www.2ndnaturewater.com/

31 AQUAOSO (2020). *Water security simplified.* https://aquaoso.com/

32 MammothWater (2020). *Water management at your fingertips.* https://mammothwater.com/products/

33 Gybe (2020). www.gybe.eco/

34 Imagine H20 (2020). *The leading ecosystem for water innovation & entrepreneurship.* www.imagineh2o.org/mission

35 Imagine H20 (2020). *Urban water challenge.* www.imagineh2o.org/urban-water-challenge

36 Ibid.

37 XPV Water Partners (2020). *Water impacts everything.* www.xpvwaterpartners.com/.

38 Smart Cover (2020). https://smartcoversystems.com/

39 Aquatic Informatics (2020). https://aquaticinformatics.com/

40 Water Finance & Management. Innovyze to be acquired by Autodesk (2021, March). https://waterfm.com/innovyze-to-be-acquired-by-autodesk/

41 Water Asset Management (2020). www.waterinv.com/home

42 W. Sarni (2021). World water day: Why now is the right time to invest in water innovation. *Nasdaq.* www.nasdaq.com/articles/world-water-day%3A-why-now-is-the-right-time-to-invest-in-water-innovation-2021-03-22

43 WaterWorld (2013, April). *The top global water research institutes.* www.waterworld.com/international/article/16208011/the-top-global-water-research-institutes

44 IHE Delft institute for Water Education (2020). www.un-ihe.org/

45 IWIDGET (2013). *iWIDGET project overview.* www.i-widget.eu/project/project-overview.html

46 Water Research Foundation (2020). www.waterrf.org/

47 K. Litwiller (2019). *Lystek thermal hydrolysis.* Lift. www.waterrf.org/news/lystek-thermal-hydrolysis

48 Lift (2019). *CNP-technology water and biosolids corp.* www.waterrf.org/news/cnp-technology-water-and-biosolids-corp#:~:text=AirPrex® is a sludge, struvite (NH4 MgPO4·6H2O).&text=In addition%2C removal of orthophosphate, polymer consumption for sludge dewatering

49 WWF (2018). *New blockchain project has potential to revolutionise seafood industry.* https://wwf.panda.org/?320232/New-Blockchain-Project-has-Potential-to-Revolutionise-Seafood-Industry

50 WWF (2019). *New technology helps WWF and partners study whales in one of the most remote places on the planet.* www.worldwildlife.org/stories/new-technology-helps-wwf-and-partners-study-whales-in-one-of-the-most-remote-places-on-the-planet

51 WWF (2020). *Generation water.* https://techhub.wwf.ca/conservation-tech-challenges/generation-water/

52 Global Newswire (2020). *WWF-Canada announces generation water tech challenge award recipients.* www.globenewswire.com/news-release/2020/01/29/1976979/0/en/WWF-Canada-announces-Generation-Water-Tech-Challenge-award-recipients.html

53 California Open Data (2020). https://data.ca.gov/

54 Digital Water (2020). www.theukwaterpartnership.org/initiatives/digital-water

55 ICT4water (2020). https://ict4water.eu/

56 AfriAlliance (2020). *About AfriAlliance.* www.afrialliance.org/about-afrialliance

57 Aqua3S (2020). https://aqua3s.eu/

58 AquaNES (2020). *The AquaNES project.* www.aquanes-h2020.eu/Default.aspx?t=1593

59 Water Innovation Europe (2020). https://waterinnovationeurope.eu/

60 Water Europe (2020). https://watereurope.eu/

5   The challenges

Thomas Edison, a renowned inventor, wisely noted that "Just because something doesn't do what you planned it to do doesn't mean it's useless." Embracing new ideas and pursuing new solutions often means there will be new challenges to address. Perhaps the greatest challenge with digital water technology transformation is that there are numerous challenges. No shortage of current and emerging issues ranging from workforce capacity to culture to cybersecurity. The good news is that digital technology start-ups, technology providers, and end users are addressing these challenges.

Despite the advancement of digital technologies and their ongoing adoption in the water sector, many challenges remain that are preventing universal digital adoption across water utilities, public sector organizations, and private sector companies. For one, a lack of support from top management along with a failure to develop a comprehensive digital strategy can be inhibitory, leaving organizations without the structure, funding, and vision necessary for successful digital adoption. Likewise, the training, financing, and security measures needed to invest in and maintain digital infrastructure can intimidate business leaders. With a well-informed business case and understanding of available resources, however, it is often clear to management that the benefits far outweigh the costs. The challenges an organization will face by pursuing digital technologies are genuine and should be thoroughly analyzed by an organization's leadership team, but they are not an insurmountable obstacle to digital adoption in public and private sector enterprises. In the following sections, several of these challenges are described in further detail along with key steps for how organizations can address them.

5.1 No strategy and/or leadership support

Potentially, the most crucial issue to discuss in relation to implementing new technologies is the business strategy (both public and private sectors) of the organization. A company without a cohesive, digital strategy has little chance of effective technology implementation and value creation. Alternatively, a digital strategy can outline a unified vision for incorporating digital technologies into business operations in a way that improves business performance and aligns

DOI: 10.4324/9780429439278-6

with company goals.[1] The successful application of a digital strategy requires it to be deeply connected with the organization's overall business strategy, with regular review of mission alignment and value creation.

Adam Tank, Director of Software Solutions at Transcend Water

Adam Tank, Director of Software Solutions at Transcend Water, believes it is critical that digital technologies create value for customers either by increasing profits or decreasing costs for customers, a belief that is echoed through Transcend's unique software offering. A start-up spinoff from Organica Water, Transcend Design Generator software creates custom preliminary designs of water facilities including related engineering documents and drawing. Designs are generated in as little as 8 <u>hours</u>, fast-tracking the time it takes for clients to choose the best design for their facility. By eliminating the usual weeks-long wait for each design iteration, Transcend enables clients to begin earning revenue sooner, reduce costs, and bring new technologies and approaches to the market faster. Adam notes, however, that value comes in many shapes and sizes and is also reflected in decreased customer complaints, resolved internal and operational issues, employee happiness, and increased win rates.

Lessons from the expert: Water is unique in terms of how it is regulated, fragmentation within the industry, and how widespread both customers and solutions are. This can make it challenging to scale technologies in the water sector. Yet, water is critical to every industry on the planet. Thus, there is a vast opportunity for digital technology providers to bring talent and innovative solutions to the water space with the potential to significantly lessen the global freshwater crisis.

It is also imperative in building a strategy that the leadership of the company is committed to the transformation initiative and understands why these advancements are critical in the short and long terms.[2] Setting the ambition at the CEO and board level increases the potential that the entire company can support the advancements taking place to their full conclusion. Pitching new technologies to the board is a critical barrier in overcoming future funding gaps, as the overall budget will likely need to be rearranged to meet the new demands (e.g., funding for technology, contracts, new IT personnel, training for current personnel on operating/utilizing digital technology, restructuring information pathways, and decision-making processes) set by a digital strategy. Likewise, goals and decisions made by top management have a larger, more direct impact on the direction an organization's digital journey will take.

Although digital projects can be pursued and implemented at lower- and mid-dle-management levels, they are most successful with the support and orchestration of an organization's C-suite and board. For example, at the Las Vegas Valley Water District, board-level adoption of digital technology goals enabled the advancement of big data and digital infrastructure to be made a priority, thereby unlocking the necessary funding for projects.[3]

One major step in securing support for digital technologies could come from developing a comprehensive digital roadmap to accompany the overall digital/business strategy. The roadmap should act as a blueprint for deploying digital technologies in the enterprise's operations, training staff, and collecting the necessary data for informing planning and decision-making. A roadmap should not only focus solely on internal operations, but, where necessary, also on the full education of end users, politicians, shareholders, management, and all staff regarding the technologies being put into place.[4] Biju George, Executive Vice President of Operations and Engineering at DC Water, states their perspective that a digital strategy and roadmap will not execute itself. It must be planned for, actively pursued, and intentionally implemented.[5] Whereas support from top management is the catalyst for accelerating an organization's digital journey, a thorough digital strategy and roadmap create the pathway enabling successful technology adoption.

5.2 Institutional silos

Many organizations continue operating in silos – business divisions that operate independently and with minimal sharing of information[6] – a common problem and critical barrier to digital technology adoption. Siloed structure can be either intentional or an accidental consequence of increasing complexity experienced by many water sector enterprises. The "silo mentality" can be horizontal or vertical, inhibiting communication and information sharing both within and between an enterprise's departments. As a result, organizations inevitably experience reduced operational efficiency and a decreased ability for managers to make informed decisions. Without comprehensive data and organizational connectivity, leaders lack the "whole picture," and decisions are left to be based on independent aspects of business operations. This lack of connectivity creates both a barrier and an opportunity for digital technologies to also transform the organizational structure and culture.

Digital systems require connectivity between departments to utilize the multiple data sets needed for meaningful operation (e.g., SCADA and AI/machine learning). Likewise, for digital technologies to be adequately deployed (per a digital strategy), there needs to be connectivity and communication between business sectors for executing technology deployment, evaluation, and the harnessing of insights and services provided by digital technologies. In addition, every company can benefit from more collaboration, alignment of vision, improved communication, trust, and accountability. The benefits are not purely in terms of building in new technology but can lead to operations savings across the entire chain. Breaking down silos is a more intensive process than it may seem at first, as it often involves redesigning processes from the top down, and even redesigning how offices are structured to encourage more cross-collaboration overall.[7] However, for the successful implementation of new technologies, silos must be eliminated.

Ways to reverse this trend include cohesive training programs to remove barriers and knowledge gaps between employees,[8] and that they motivate and incentivize executives and management teams toward achieving these goals. In the long run, avoiding dealing with silos will prove detrimental to the overall company's health and toward competitiveness.[9]

5.3 Energy-intensive training

Oftentimes, there will be a significant learning curve for employees on using the new technologies that the organization has committed to implement. Individuals in operating, management, and decision-making roles will need to understand how to integrate, run, monitor, and troubleshoot new technologies as well as interpret data or insights provided by digital systems. In some cases, new IT staff or engineers may need to be brought on board who have a stronger background in digital technologies. Some executive management may worry that the costs of training or expanding their staff outweighs the benefits. However, this is not the case when your company wants to get ahead and focus on a long-lasting culture of innovation that can realize immediate competitive benefits.

Training can typically be structured in efficient and effective ways, such as degree and certificate programs, in-person training courses, and online learning. Likewise, there are informal conferences that employees can attend or take advantage of free online resources such as TED talks, podcasts, webcasts, and articles. The number of resources available for initial training is substantial and not limited to these, there can also be external collaborations and projects leading to new experiential knowledge.

Public and private sector enterprises should understand that in this process, they do not have to go it alone. Multiple enterprises can collaborate for training and knowledge sharing, as well as learn from those that have already undertaken the process. Likewise, a company can search in its local community for accelerators and tech hubs that may offer resources for fully operating and understanding the technologies at hand. This could also potentially lead to new connections which facilitate access to experts and other technologies that the organization might require later in the digital transformation process.[10] In addition to focusing on the overall advancement of employees, companies should also focus on developing leaders within their company who can be digital leaders for the sector as a whole. These leaders can travel and exchange ideas between technology companies and technology consumers (including their own organization), thus accelerating the pace of adoption and innovation. Every interaction can make a difference when it comes to the scale of changes that will be necessary going forward.

5.4 Proper financing

There will always be a desire by water users to reduce costs. Especially now, as the COVID-19 pandemic and its ensuing economic downturn leaves many consumers and companies alike in financial distress, raising prices to cover costs is not a sustainable solution for technology uptake in most areas. At least when communicating with customers and gaining the support of top management, it is important to distinguish that many of the new technologies will reduce costs over time through lower energy consumption, chemical use, and labor costs. This is also an incentive for promoting the capital investments necessary for most projects.[11] Likewise, more efficient, enhanced operations along with the data insights provided by digital technologies play a major role in both value creation and risk mitigation. Yet, when it comes to investments, senior leadership is often more inclined to fund projects framed as contributing to business growth. Therefore, pitching digital investments as a business opportunity and pathway to long-term savings can help organizations break down the funding barrier.

One way to explore fit and efficacy of digital solutions before making a large, financial commitment is to pilot a technology first. Pilot projects offer a cheaper way to verify that the investment your organization wants to make is the correct one for the given context. At the conclusion of a pilot project, the value case of that specific technology needs to be crystal clear for the company to move forward with further investment.[12] Successful pilot projects offer both an opportunity for the technology company to refine their product and for the host company to prepare for a larger-scale technology rollout.

Clay Kraus, VP of Revenue at Flume

Clay Kraus has worked in business and product development for much of his career but most recently for Rachio (smart controller systems for at-home irrigation) and now Flume (smart home water monitor for leak detection and water use analytics). Both companies bring IoT technology and their accompanying software platforms to homeowners for informed, real-time water management/monitoring. At Rachio, a core element of successful projects has been the company's unique, cross-sector partnerships with water utilities. Through his position, Clay worked with utilities to develop subsidy programs for Rachio water controllers, thus creating a dynamic that is ultimately a comarketing arrangement designed to achieve reductions in water use. Such public–private partnerships create value for Rachio, the utility, and homeowners alike: Rachio gains a larger customer base and experiences accelerated technology adoption; utilities avoid costs associated with infrastructure, developing new supply, and customer support; and homeowners reap savings on their water bill.

With proven success at Rachio, Clay is now working to replicate this business model at Flume.

Lessons from the expert: As Clay has shown through his experiences at Rachio and Flume, exploring new business models (e.g., public–private partnerships) and funding mechanisms (e.g., subsidy programs) can help advance digital technology adoption and overcome competitive barriers for new technology companies.

There are several additional approaches which can make investing in digital water technologies more attractive. Companies need to first ensure they are considering the range of risks involved in their new technology initiative and that it will in fact lead to a more stable revenue stream than before. Utilizing blended finance is also an approach used to unlock further financing sources and improve the risk-return profile of investments, particularly in developing countries.[13] The OECD has begun work on creating *Principles for Blended Finance* for water which will present more specific and actionable recommendations. Nonetheless, it is clear that funding remains a significant hurdle for companies to overcome regarding digital technology investments. More growth and improvement are needed around finance opportunities for technology investments to move the water sector forward.

5.5 The true cost of water

Part of the funding gap regarding public and private enterprises' digital water projects stems from a failure by most organizations to recognize the true cost of water. Although essential for life – and becoming increasingly scarce – the value of water is underappreciated in the market.[14] As Rebekah Eggers, then IBM's WW Leader on IoT for Energy, Environment, and Utilities, noted in a 2018 interview, most societies consider water a cheap commodity and, in many places, a right.[15] The perceived low cost of water has two primary effects. First, enterprises that have access to low-cost water lack an incentive to pursue water-saving projects (e.g., digital technologies).[16] In other words, so long as water remains cheap, organizations will have little interest in spending money to improve efficiency or streamline processes. Second, the failure to recognize the true cost of water at the enterprise level is reflected in the cost end users pay for products and services. As a result, organizations fail to raise the funds necessary to finance investments in digital projects.

This phenomenon is especially relevant in the US water utility sector. Burdened by aging infrastructure, nonrevenue water, and maintaining source water conditions – challenges where digital technologies could play a major role in early problem identification and prioritization as well as reducing costs and improving efficiency – the water utility sector faces enormous funding gaps due in part by the disparity between the true versus imposed cost of water. Turning to other regions, in countries such as Israel where the true cost of water is incorporated into water rates, utilities have had adequate funding for massive utility projects and water infrastructure expansion.[17] Recognizing the true value of water has led Israel to become a global leader in both the development and extensive use of water sector innovations, including digital technologies. Moving forward, funding and other investment incentives will continue to be a barrier to digital water projects unless public and private sector enterprises more widely acknowledge the true cost of water.

5.6 Cybersecurity

On February 5, 2021, a hacker gained control of the Oldsmar, Florida, water treatment facility and attempted to increase the amount of sodium hydroxide in the water purification process.[18] Social media and the press were buzzing about the story, and some water sector professionals engaged in the perils of adopting digital technologies as if the choice was analog versus digital. While a bit sensational, it is really the wrong question.

The real issue is that water facilities struggle to adequately invest in effective cybersecurity solutions even when confronting "an increase in the frequency, diversity, and complexity of cyberthreats," as a 2020 study of 15 cybersecurity incidents in the water sector found.[19] A 2020 survey found that just 19 percent of water utilities said they were confident that their rates and fees could cover the cost of existing services, never mind pursuing infrastructure upgrades. While there are some notable exceptions, many water utilities in the United States are small and saddled with aging infrastructure. They may "only have one or two, maybe three, IT folks who manage the network," said Chris Sistrunk, a technical manager at Mandiant, the incident response arm of security firm FireEye.

The incident in Florida confirms concerns by public and private sector enterprises and society at large about the security of their data and the security of their machine operations. As technology advances, so does the sophistication of hackers. Outsider malicious actors gain access through a simple phishing email or an unprotected portal intended for a trusted equipment manufacturer or service technician access. Trusted guests, such as the equipment manufacturer and/or service technicians, could be the victim of an outsider threat due to their own cybersecurity weakness and may be targeted unknowingly to serve as a carrier. The data payload they exchange with machine control systems could be used as a trojan horse to inject out of range operating commands. Insider malicious actors are given access or have the credentials to create access certificates beyond their assigned scope, role, or permitted actions and then act anywhere in the network under an alias.

There are several factors that make public and private enterprises and public utilities vulnerable. Within the public utility sector, water utilities, in particular, are especially at risk because they are geographically distributed, diverse in scale of service, often operate independently, and work within very tight budgets.

These factors add up to limit available investment in human capital and technology associated with cybersecurity. As a result, water utilities become exploitable due to insufficient cyber hygiene (Figure 5.1)[20] which provides identity management, access control, malware detection; minimal network segmentation that compartmentalizes permitted user machine actions and functional commands; limited network surveillance for detection of abnormal user patterns and isolates critical machines; and limited machine-level security tools that ensure operating parameter compliance.

The situation is compounded as manufacturers increasingly create and distribute machines that must be connected to achieve the operational effectiveness, production cost efficiency, machine availability, and maintenance cost reduction required to generate the expected return on investment. The pitfall is that these benefits come with risks such as exploitable service access, failure of operators to change vendor default settings, failure to maintain firmware and operating system currency, failure to remove or update antiquated equipment, and overlooking the network when assessing risks.

Beyond the technical risks that the utility faces, there are also reputational, regulatory, and civil liability risks. To express the severity of these risks, most states in the United States have legislation imposing criminal penalties for the failure to disclose data breaches in time.[21] This leaves serious consequences for those who fail to protect themselves and their customers' data. For larger organizations and utilities, particularly those more reliant on digital technologies and automated processes, it can be a daunting task to prepare for this

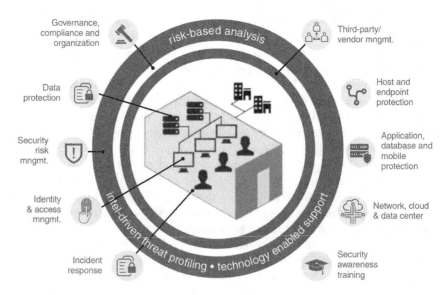

Figure 5.1 Critical Components of a Cybersecurity Program

Source: Adapted from McLellan (2018).

reality.[22] Two initial responses to the threat of data breaches are anonymizing customer data and implementing cybersecurity systems alongside new technologies. However, vigilance and constant updating are still necessary to keep digital systems up to date and avoid unnecessary risks.[23]

There are certain frameworks that can help businesses prepare for cybersecurity risks. These include the National Institute of Standards and Technology (NIST) cybersecurity framework, which is a voluntary set of standards, guidelines, and best practices for managing cyber risk.[24] There is also the American Water Works Association's *Process Control System Security Guidance for the Water Sector* that comes with a corresponding Use-Case Tool for establishing and improving cybersecurity systems specific for utility operations technology as well as enterprise security practices.[25] The tool generates recommended controls based on the individual characteristics of that water utility. Moreover, it is prudent for organizations to take out cyber insurance on their data and systems. Finding the right plan can be difficult due to the complexity of the systems that are being covered; however, more are becoming customized to meet organizations' needs.[26]

Beyond the plant – cybersecurity in the digital water enterprise

Thus far, the focus of the discussion has been on cybersecurity as it relates to "inside the plant" operations. This section of the discussion will focus on the rapidly emerging digital water enterprise taking place at the management level in functions such as customer account management, water delivery management, wastewater collection management, and asset management. These functions are increasingly migrating from legacy in-house models to as-a-service models where digital water service companies use local data to set up billing and process payments; provide water delivery and wastewater collection management dashboards; and support asset management dashboards and tools for leak detection, preventive maintenance, and prioritization of capital improvement projects. These services have proven themselves in small, medium, and large utilities and are increasingly becoming part of the portfolio of citizen-focused digital transformation projects being implemented in "Smart Cities" worldwide.

While every municipality and utility will plan their own digital services journey, they will also need to plan their cybersecurity journey. As an example, a choice to use a cloud-based asset management system sets in motion the need to adapt to a distributed network environment that will likely include Internet of Things sensors that detect pressure changes. These sensors will transmit data via radio, cellular services, or fiber to the cloud-hosted leak detection application, which will then provide monitoring dashboards on desktops, laptops, tablets, and phones.

In this new environment, traditional cybersecurity models are disrupted – those that were associated with fixed physical networks and fixed interfacing devices such as hardwired desktop computers or machine terminals relied on perimeter for protection. The principle was that the interface device was known and trusted, the network was known and trusted, and the user signed-in and became trusted. Once trusted, the user was permitted to conduct interactive sessions inside the perimeter with enterprise applications such as email, enterprise data systems, and operational technology such as machine controllers. This concept of "once in, always trusted" is now challenged in a world where the interface device can vary throughout the day from mobile to desktop, the network can vary throughout the day from cellular to public Wi-Fi to private Wi-Fi or fixed connection, and a "user" can now be person, a cloud application requesting a data update from a local server, or a sensor in the field sending a data burst.

In effect, the network just became an ever-evolving constellation of network of users, interface devices, and networks of origin. In response, NIST moved away from perimeter-centric to identity-centric cybersecurity in an approach known as "Zero Trust."

Fundamentally, Zero Trust changes two major aspects of access and network navigation – a level of trust *must be explicitly established* and that level of trust *applies only for the transaction requested.*[27] Once the transaction is complete, the trust expires and further transactions will require reauthentication. In a zero-trust environment, every user, device, and network become a traceable element of access control and trust.

Cybersecurity has become more top of mind for water and wastewater utilities over the past year as adoption of digital technologies takes hold, remote working increases and incidents of cyber security breaches increase. Cybersecurity as an essential component of water infrastructure and, as a result, is now being addressed by both hardware and software water technology companies.

5.7 Data integration

One challenge to consider is that every public and private enterprise worldwide is starting at a different level of technological capacity; therefore, the range and quality of data each organization has varies widely too. Nonetheless, all organizations will need data collection that is reliable and that starts with sensors. A current barrier to widespread deployment of sensors is their cost, however, that seems to be falling each year as the technology advances.[28] The next challenge will be what to do with the amount of data coming in from various sensors and how to integrate new types of data for advanced uses such as AI, blockchain, and their related applications. Enterprises will need to have undergone at least the strategizing and breaking down silos steps mentioned earlier to successfully be able to process data from across their entire operations into useful business insights and predictive analytics. As the water industry is sometimes described as "data rich and information poor," it is imperative for the capacity of the industry to step up to make the digital transformation a desirable reality.[29]

Easing this transition is the growing sophistication of data management and analysis software available, where software solutions are becoming more oriented toward specific applications such as leak detection, meter management, and condition monitoring. For example, IBM's Watson Analytics incorporates complex computing and advanced analytics to generate high-value insights in a less complicated manner than before. More advanced software means less work needs to be done on the part of the organization to collect and analyze data. Everything needed – from data collection to providing data insights – is managed directly through the software.

Likewise, the rise of cloud computing is quickly becoming a more economical and flexible option for data management.[30] Cloud computing makes it easier for organizations to manage large data sets and integrate data from multiple locations, processes, and levels of operation. Cloud systems are also able to be accessed and managed off-site, which provides necessary reliability and accessibility to data and operations in the field and from multiple locations. Companies that currently have more traditional or outdated technology infrastructure will have a larger hurdle to cross when it comes to integrating data and establishing digital systems. With the increased availability and capabilities of advanced data integration systems, however, the hurdle is becoming ever easier to surmount.

5.8 Scale-up challenges

For the most recent emerging technology companies, specifically AI, blockchain, and new software, as well as hardware such as advanced sensors, meters, and drones – there is an adjustment period between when the technology is first launched and when the company itself is mature enough to deploy the technology at a large scale. Issues with small start-ups teams can arise such as shortages in labor, less service available for maintenance, and cultural mismatches with large corporate clients. Several large-scale companies which act as technology incubators in the water sector, including Suez and Xylem Inc., offer start-ups the supporting infrastructure to succeed across the entire industry. However, collaborating with such companies does not always eliminate the challenges that a starting technology company may face. Therefore, it is imperative for digital technology providers to understand these challenges and strive for greater mutually beneficial support between them and their clients. Further challenges that a start-up may face include product awareness and visibility among potential clientele; lack of proven use cases; technical complexity of products; barriers from regulation and policy; and the complex, multilevel decision-making dynamic of customer organizations.[31] It is also encouraged that technology experts be consulted and on hand during implementation of new, digital technologies and afterward to ensure a smoother transition.

Within the organization adopting digital technologies, scale can again be a challenge. For public and private enterprises that already have a digital foundation, new projects build on basic systems and staff are already adept at delivering change.[32] As data and expertise accumulate, taking on new projects becomes easier and easier. Conversely, it can be difficult for organizations in "digital poverty" – those without a robust, digital baseline – to initiate digital projects.

A commitment to training and adherence to the organization's digital roadmap as described previously can help the organization escape digital poverty and unleash the benefits digital technologies provide.

5.9 Ease of use

Another challenge corresponding to the training of employees is the adaptability of those employees toward using the new technologies. A strong consideration of the technology providers should be the ease of use of their product and how quickly someone can be trained to use it. The higher the efficiency of these two factors, the higher the likelihood of widespread adoption by technology consumers. Currently, dependent upon the baseline capacity of an individual organization, most employees will not be automatically ready to start using new hardware and software as it will often require a significant deviance from the current technology in use. An example being sensors, where digital is rapidly replacing analog, there is still a learning curve. New technology needs to be adding value to the organization's operations, which can thereby justify the retraining necessary and encourage higher adoption rates from the beginning.[33] Ultimately, digital technologies that are unnecessarily complex for the end user will not be adopted or will be incorporated but either incorrectly or altogether unused, depriving organizations of critical data insights.

5.10 Knowing/doing gap

While it is easy to write recommendations that public and private enterprises should use to overcome these challenges, there is still a real gap between knowing and doing. Companies are encouraged to build roadmaps and share knowledge as outlined previously; however, there are certain ways that mature organizations tackle these challenges. Described by Kane *et al.* (2019) in *The Technology Fallacy* as building a company's "digital DNA," bridging the knowing–doing gap requires that an organization create a culture of change and have leaders with the skills and agile mindset needed for adapting to digital technologies (Figure 5.2). Organizations must be willing to identify and assimilate new innovations; understand the strategic value technologies can unfold over time; implement a test, learn, and scale digital models internally; and most importantly, attract and retain strong, digital employees and leaders. This all requires that a company put forward the effort to expand their talent search beyond the usual suspects and that they commit fully to employee learning and development.[34]

To implement the "doing" part, it is also important to take the following actions within a company. First, it is useful to model the behaviors that an employee wants to see to encourage development of the organization's culture. Second, the sensing and discovery capacity of the organization must be improved. Likewise, regular testing to develop the business case should be incorporated for each new technology. Evaluations to deduce lessons from each

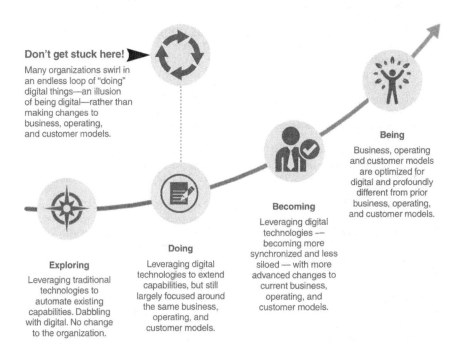

Don't get stuck here!

Many organizations swirl in an endless loop of "doing" digital things—an illusion of being digital—rather than making changes to business, operating, and customer models.

Being

Business, operating and customer models are optimized for digital and profoundly different from prior business, operating, and customer models.

Becoming

Leveraging digital technologies — becoming more synchronized and less siloed — with more advanced changes to current business, operating, and customer models.

Doing

Leveraging digital technologies to extend capabilities, but still largely focused around the same business, operating, and customer models.

Exploring

Leveraging traditional technologies to automate existing capabilities. Dabbling with digital. No change to the organization.

Figure 5.2 The Knowing Doing Gap

Source: Adapted from Kane *et al.* (2019).

test must be collected, properly synthesized, and communicated to ensure that the effort in testing is worthwhile and aligned with business goals. Finally, tests should evaluate the technology's scalability and how the job functions of employees would need to evolve to incorporate the new technology. This can enable a company to begin to close the knowing–doing gap and demonstrate measurable progress against their own business goals as well as on their digital transformation journey.[35]

Notes

1 C. Reyes (2020). *What is digital strategy?* Liferay. www.liferay.com/resources/l/digital-strategy#:~:text=Digital%20strategy%20focuses%20on%20using,use%20to%20achieve%20these%20changes
2 W. Sarni, C. White, R. Webb, K. Cross, & R. Glotzbach (2019). *Digital water: Industry leaders chart the transformation journey.* The International Water Association.
3 Ibid.
4 Ibid.
5 Ibid.
6 W. Kenton (2020). *Silo mentality.* Investopedia. www.investopedia.com/terms/s/silo-mentality.asp#:~:text=In%20business%2C%20organizational%20silos%20refer,shared%20because%20of%20system%20limitations

7 B. Gleeson (2017). *5 Ways to destroy the pesky silos in your organization*. Inc. www.inc.com/brent-gleeson/5-ways-to-destroy-the-pesky-silos-in-your-organization.html

8 G. C. Kane, A. Phillips, J. Copulsky, & G. Andrus (2019). *The technology fallacy: How people are the real key to digital transformation*. MIT Press.

9 B. Gleeson (2017). *5 Ways to destroy the pesky silos in your organization*. Inc. www.inc.com/brent-gleeson/5-ways-to-destroy-the-pesky-silos-in-your-organization.html

10 W. Sarni, C. White, R. Webb, K. Cross, & R. Glotzbach (2019). *Digital water: Industry leaders chart the transformation journey*. The International Water Association.

11 W. Sarni & H. Share (2019). *From corporate water risk to value creation*. Global Water Intelligence. www.globalwaterintel.com/news/2019/31/from-corporate-water-risk-to-value-creation

12 W. Sarni, C. White, R. Webb, K. Cross, & R. Glotzbach (2019). *Digital water: Industry leaders chart the transformation journey*. The International Water Association.

13 OECD (2018). Financing water: Investing in sustainable growth. *OECD Environment Policy Papers*, 11(11), 16.

14 M. De Stefano (2019, March). The digital transformation of water. *Smart Water Magazine*. https://smartwatermagazine.com/blogs/maurizio-de-stefano/digital-transformation-water

15 M. Jimenez (2018). *The impact of digitalisation on the water sector – An interview with Rebekah eggers*. International Water Association. https://iwa-network.org/the-real-impact-of-digitalisation-on-the-water-sector/

16 Ecolab (2019). *Knowing the true cost of water: Industrial water solutions at Singapore international water week*. https://en-au.ecolab.com/news/2019/06/knowing-the-true-cost-of-water

17 S. M. Siegel (2015). *Let there be water*. St. Martin's Press.

18 A. Hassanzadeh, A. Rasekh, S. Galelli, M. Aghashahi, R. Taormina, A. Ostfeld, & M. K. Banks (2020). A review of cybersecurity incidents in the water sector. *Journal of Environmental Engineering*, 146(15). https://doi.org/10.1061/(asce)ee.1943-7870.0001686

19 Ibid.

20 C. McLellan (2018). Cybersecurity: How to devise a winning strategy. *ZDNet*. www.zdnet.com/article/cybersecurity-how-to-devise-a-winning-strategy/.

21 N. Ramirez (2020). *The great big list of data privacy laws by state*. Osano. www.osano.com/articles/data-privacy-laws-by-state

22 J. H. Germano (2019). *Cybersecurity risk & responsibility in the water sector*. American Water Works Association. www.awwa.org/Portals/0/AWWA/Government/AWWA-CybersecurityRiskandResponsibility.pdf?ver=2018-12-05-123319-013

23 W. Sarni, C. White, R. Webb, K. Cross, & R. Glotzbach (2019). *Digital water: Industry leaders chart the transformation journey*. The International Water Association.

24 M. Barrett (2018). Framework for improving critical infrastructure cybersecurity. *Proceedings of the Annual ISA Analysis Division Symposium*, 535, 9–25.

25 P. Gaberdiel & K. Morley (2015). Process control system security guidance for the water sector. *Proceedings of the Water Environment Federation*, 2014(1), 1–5.

26 J. H. Germano (2019). *Cybersecurity risk & responsibility in the water sector*. American Water Works Association. www.awwa.org/Portals/0/AWWA/Government/AWWA-CybersecurityRiskandResponsibility.pdf?ver=2018-12-05-123319-013

27 S. W. Rose, O. Borchert, S. Mitchell, & S. Connelly (2020). *Zero trust architecture*. National Institute of Standards and Technology. https://doi.org/https://doi.org/10.6028/NIST.SP.800-207

28 B. Sivathanu (2019). Adoption of industrial IoT (IIoT) in auto-component manufacturing SMEs in India. *Information Resources Management Journal*, 32(2), 52–75. https://doi.org/10.4018/IRMJ.2019040103

29 C. Brzozowski (2018, November). Accessibility and analysis. *WaterWorld*. www.waterworld.com/home/article/14070973/accessibility-and-analysis

30 Ibid.

31 M. Zwilling (2018). *5 Barriers to business growth that new ventures rarely anticipate.* Inc. www.inc.com/martin-zwilling/5-barriers-to-business-growth-that-new-ventures-rarely-anticipate.html
32 Aqua Tech (2019, December 6). *Five barriers preventing a digital water utility.* www.aquatechtrade.com/news/utilities/digital-water-utility-barriers/
33 K. Regn (2019). *What will AI really do for the water industry?* Raconteur. www.raconteur.net/technology/artificial-intelligence/water-technology-ai/
34 G. C. Kane, A. Phillips, J. Copulsky, & G. Andrus (2019). *The technology fallacy: How people are the real key to digital transformation.* MIT Press.
35 Ibid.

6 The roadmap

What will technology look like ten years from now? Or 20? How about 50 years from now? There is no way to know for sure, but as we're well into the Fourth Industrial Revolution it's safe to say there's no turning back. Digital technologies will continue evolving in years to come – emerging with new capabilities and able to address new problems – and the technologies available today will be a necessary component of their foundation. Little is known about when and how change will occur, just that, as Ernest Hemingway says in *The Sun Also Rises*, disruption occurs "gradually and then suddenly." Enterprises that not only fail to adopt digital technology but also inadequately prepare for the systemic changes in operations, strategy, and business ecosystems that stem from the digital transformation will get left behind. In the water sector – already notorious for being a late-stage adopter – getting left behind is not an option. Water is not only a human right, but also it is necessary for all things, from organisms and ecosystems to industries and civilizations, to function. As climate change, population growth, and urbanization, among other factors, continue to place stress on already limited resources, it is critical that the water sector embraces the digital technology wave.

By harnessing digital water technologies, public and private sector enterprises can unlock better water management, universal access to safe drinking water, economic growth, and thriving ecosystems. The potential for Fourth Industrial Revolution technologies to solve some of the world's biggest problems is enormous, but that potential is not yet being met.[1] According to the World Economic Forum (WEF), as of fall 2020, we are off track to meet the Sustainable Development Goals by 2030. Regarding SDG 6 specifically, no global region is on track to achieve universal basic sanitation access by 2030.[2] Yet, there is enormous opportunity for digital technologies to accelerate progress against the SDGs all while stimulating economic growth, creating jobs, and ensuring a more resilient, secure, and equitable water future (Figure 6.1).

In terms of economic gains, the WEF estimates achieving the SDGs would generate $12 trillion a year in revenue and cost savings and create an additional

DOI: 10.4324/9780429439278-7

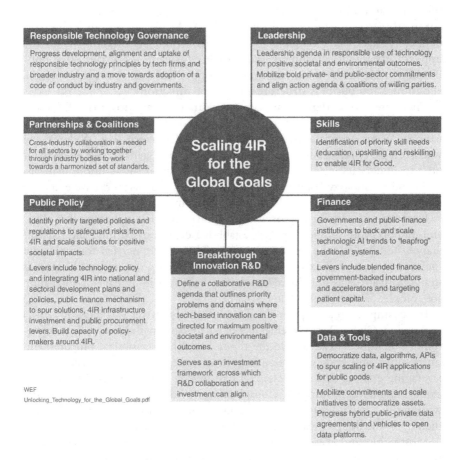

Figure 6.1 Key Enablers to Harness Technology for the SDGs

Source: Adapted from World Economic Forum (2020).

380 million new jobs by 2030.[3] Likewise, the technology sector alone has the potential to generate an additional $2.1 trillion annually by applying digital solutions to the global challenges. Aside from the SDGs, in 2018, an estimated $1.2 trillion was spent by companies on their own digital transformation.[4] However, the current opportunity is much larger than profits and interorganizational transformation.

The COVID-19 pandemic opened the door to rapid, wide-spread changes in operations, intraorganizational structure, and relationships and opened minds to new ways of doing business.[5] It exposed long-standing societal disparities, especially regarding access to digital technologies and healthcare. By disrupting business and societal norms, the pandemic created the space and illuminated the need for rapid digitalization (e.g., increased automation, connectivity, and reliable data). As we rebuild after the pandemic, reestablishing prepandemic systems would be antagonistic to the opportunity the disruption created. Moving forward, it is critical that public and private sector enterprises use this opportunity to reimagine societal, economic, and business structure and future directions.

There are several digital transformation strategies and roadmaps for water in both the public and private sectors. Water industry associations (e.g., International Water Association), technology solution providers (e.g., Xylem), and water research organizations (e.g., Global Water Intelligence) have all contributed to understanding digital transformation for water. In addition, management and technology companies such as Deloitte[6] and Gartner[7] have extensive bodies of work and processes for digital transformation.

In many respects, the approaches are similar: develop a strategy that aligns and supports your business strategy, establish a culture of learning and risk tolerance, and provide the human and financial resources needed to implement the strategy, pilot and experiment and scale. While on the surface this sounds straightforward, it is not. In addition, the challenges in digital transformation are complex as noted in Chapter 5.

For example, a roadmap for the digital future for water was recently (2021) outlined by Global Water Intelligence in partnership with Nokia.[8] Key opportunities identified were around the themes of data accessibility, procurement, smart cities, and best practices. Notably, investing in data infrastructure and leveraging existing data is an essential first step as various enterprises begin to explore digital opportunities. Consolidating existing digital technologies and working to unlock legacy systems through connectivity and data sharing can then open the door to realizing the full benefits of digital solutions.

Most public and private sector enterprises have large amounts of data which is often siloed due to legacy systems. Since digital tools (e.g., AI) are generally limited by the quality of their input data, such fragmentation prevents organizations from fully utilizing data for informed decision-making. Service-based software options and migration to the cloud can help eliminate data silos and improve access to and the quality of an organization's data. With accessible data, powerful tools then become available (e.g., machine learning for pattern recognition, digital twins, and network simulation). In addition, dashboards from software services can help visualize key insights and track an organization's progress against KPIs. A strong foundation in data collection and infrastructure is therefore necessary to realizing the benefits of digital technologies.

Beyond unlocking legacy systems, data should also be centralized, integrated, and incorporate common standards at the enterprise level. More so, enterprises must facilitate the company-wide culture shifts and buy-in needed to make digital projects successful. Additional priorities should include "data literacy in training and recruitment, democratizing access to data across the organization, and fostering innovation through interdepartmental collaboration."

As organizations work to integrate digital technologies across their company, it is important they remember that the burden associated with new technologies may fall more heavily on some departments than others. Having open communication and ensuring the business case goes beyond ROI to explain value added in employees' everyday jobs can ensure smoother transitions to digital technology.

Procurement of digital technologies poses another opportunity (and challenge) to the digitalization of public and private enterprises. Organizations must understand and be able to clearly articulate their technological and cultural readiness in a request for proposal. Likewise, vendors must clearly understand client organizations' capabilities and pain points and focus on tailored services rather than a one-size-fits-all approach. Here, it becomes critical that organizations have quality, accessible data. Preexisting data enable vendors to utilize relevant information to pitch customized solutions. In a data-driven world, with the needs of an organization changing just as rapidly as the technology solutions available to them, it is especially important that vendors and clients work more as partners than transactional parties. Digital technology providers and their clients work together to codevelop solutions that are unique for each organizations' baseline conditions (Figure 6.2).

Figure 6.2 Risks and Rewards of Client–Provider Partnerships

Source: Adapted from Gould and Weaver (2021).

Additionally, utilizing performance-based contracts can help balance the risks and rewards associated with investing in new technology or software.

As public and private enterprises continue to digitalize, benefits will be realized not just at the organizational level, but at local and regional levels as well. Increased connectivity and the integration of data sets will allow broader trends to be realized and provide insights on external water management challenges. For example, digital technologies are further enabling smart cities through IoT and data sharing. In a smart city, connectivity between water and wastewater treatment utilities, underground infrastructure, source/effluent water quality monitoring, and customer meters can enable real-time response to customer demands, changing water conditions, and pipe breaks or leaks. Such interconnectivity and data sharing not only streamline processes but also enable holistic planning and improved quality of life. Stakeholder engagement – especially with citizens – will be central to the successful development of smart cities by ensuring awareness and buy-in among residents.

At the enterprise level, the digital transformation journey is unlikely to be the same for any two water sector organizations and will require a variety of strategic, intentional actions. The key elements of a digital water transformation strategy are provided in the following, and a few "guiding principles are illustrated in Figure 6.3.[9]

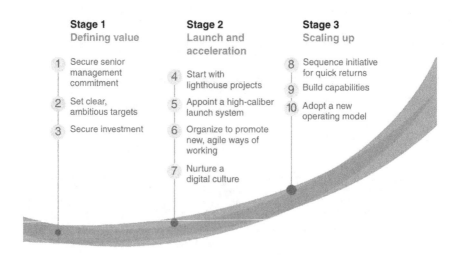

Figure 6.3 The Ten Guiding Principles of Digital Transformation

Source: Adapted from Tanguy *et al.* (2017).

- *Develop a digital transformation strategy that aligns with and supports your business strategy*
This sounds obvious. Unless a digital transformation strategy supports your business objectives, there is no point in investing human capital and money in it. Similar to when companies develop and scale successful sustainability strategies, they ensure business value is created and measured. This alignment with business strategy requires senior level leadership and commitment to maximize the potential for success.
- *Build a culture of learning and innovation and risk taking*
As previously discussed, a culture that truly supports a digital transformation strategy is critical. As discussed in, *The Technology Fallacy*, culture will make or break a digital technology adoption strategy. The culture must value learning from both inside and outside the organization and support risk taking. The status quo is the greatest barrier to innovation and does not change easily in a culture where risk is not encouraged.
- *Build the digital workforce*
The workforce is the most important aspect of successfully adopting and scaling a digital strategy. This translates to investment in training and the tools needed to succeed. This is not something to do "on the cheap." No amount of innovative technologies will deliver value unless the workforce has the capacity and tools to ensure successful adoption. Investing in the digital workforce pays dividends in terms of recruitment and retention, again from *The Technology Fallacy*.
- *Commit to aligning ambition with SDG best practices*
First and foremost, all public and private sector enterprises should commit to aligning their organization's ambition with the SDGs. Commitments to action must be communicated from a board/executive level yet in a way that encourages enterprise-wide, meaningful participation. It is especially important that commitments align with company culture and stakeholder values to ensure longevity and success of both the organization and any set goals (e.g., science-based targets).
- *Embed SDGs and digital transformation efforts in core strategy*
Transformation occurs when new ideas, processes, and technologies are embedded throughout the organization rather than being siloed as a side project. Of course, smaller pilot tests should be run to gather lessons from and determine the success of digital projects before scaling, but consequential change ultimately happens when digital and sustainability goals are woven into the organization's core strategy.
- *Develop targets and KPIs to hold leadership and the organization accountable*
Key performance indicators can keep an organization focused and on track with their digital strategy as they work to achieve both their company and sustainability (although the two should be one and the same) goals. Measuring and reporting mechanisms are important to hold the organization accountable for the responsible and strategic use of technology toward meeting the SDGs.

- *Develop tech-enabled products and services*
  Product development teams should consider how to minimize the environmental footprint of their products by using digital technologies. Likewise, organizations should consider how to digitalize their product or service offerings to remain relevant and competitive with their customers/ clients as the water sector continues to digitalize.
- *Enhance your organization's talent pool*
  One of the biggest challenges organizations will face in the coming years is attracting and retaining talent. Incorporating and operating digital technologies, as well as working with the data and software that go with them, will require new and highly sought-after skills. Water sector enterprises will need to assess skill gaps in their organization, commit to recruiting skills they cannot build, and provide opportunities for employees to develop new or build on current skills. Skills are constantly changing in the digital age and some become obsolete in a matter of years, meaning organizations will need steadfast commitment to employee growth, talent retention, and acquisition.
- *Engage with policymakers*
  Public and private sector enterprises need agile and informed policies that are consistent with evolving technologies. In many cases, technology policies lag well behind the technologies which they regulate, yet regulations provide necessary structure and guidance for tech users and developers alike.[10] Public and private sector enterprises should have a public engagement strategy and be intentionally involved with developing digital technology policy as well as policies that advance progress toward the SDGs.
- *Establish multistakeholder and cross-sector partnerships*
  Multistakeholder and cross-sector partnerships are both a driver and a result of the ongoing digital transformation of the water sector. Such partnerships enable public and private enterprises to combine resources and expertise in the research, development, and deployment of digital technologies for solving water challenges. Likewise, they enable knowledge sharing so that organizations can learn from the digital water journey of others. By collaborating across the ecosystem of water and beyond, water sector enterprises can accelerate their own digital transformation, advance innovation and industry knowledge, and enable better water management for all.

As the WEF puts it, this is the decade of action. Traditional policy and market responses are not sufficient to achieve the global goals around water scarcity, security, and equity. In addition, adoption of digital water technologies remains patchy – largely resulting from disparities in enabling conditions (e.g., electricity, connectivity, and infrastructure). Achieving the SDGs will require deploying digital technologies at scale in the water sector, meaning such disparities must be addressed first followed by extensive cross-ecosystem collaboration and innovation. As a wicked problem, water is fundamentally interconnected with numerous other societal challenges such as poverty, equality, and women's rights to name a few. Addressing these challenges together will be the key to solving water.

Notes

1 C. Herweijer, B. Combes, & A. Gawel (2020). *Harnessing technology for the global goals: A framework for corporate action.* World Economic Forum.
2 2030 Vision (2017). *Uniting to deliver technology for the global goals.* www.2030vision.com/news/2030vision-uniting-to-deliver-technology-for-the-global-goals
3 Ibid.
4 World Economic Forum (2020). *Unlocking technology for the global goals.* World Economic Forum. http://www3.weforum.org/docs/Unlocking_Technology_for_the_Global_Goals.pdf
5 C. Herweijer, B. Combes, & A. Gawel (2020). *Harnessing technology for the global goals: A framework for corporate action.* World Economic Forum.
6 Digital Transformation (2021). *Deloitte.* https://www2.deloitte.com/za/en/pages/digital/topics/digital-transformation.html.
7 The IT Roadmap for Digital Business Transformation (2020). *Gartner, Inc.* https://emtemp.gcom.cloud/ngw/globalassets/en/information-technology/documents/insights/the-gartner-it-roadmap-for-digital-buisness-transformation-excerpt.pdf
8 T. Gould & R. Weaver (2021). *Digital futures: Creating a roadmap for utility performance.* Global Water Intelligence.
9 C. Tanguy, J. T. Lorenz, B. Sternfels, & P. Willmott (2017). *A roadmap for a digital transformation.* McKinsey & Company. www.mckinsey.com/industries/financial-services/our-insights/a-roadmap-for-a-digital-transformation#
10 G. C. Kane, A. Phillips, J. Copulsky, & G. Andrus (2019). *The technology fallacy: How people are the real key to digital transformation.* MIT Press.

Closing

Ivan Lalovic, Founder and CEO of Gybe

The long and winding road of hardware development and digital adoption

Digital technologies are transforming how we live and work. The adoption of digital devices, services, and information technologies is global and has caused sweeping changes to almost all spheres of human endeavor. From communication between individuals, to news and mass media, to entertainment, the changes caused by an individual's increased ability to generate digital content and share their point of view are rippling through social, political, and business landscapes. Although much of the general public thinks of the current advancements as rapid leaps forward, the digital era – exemplified by mobile computing, cloud memory storage, and wireless broadband networking – is the culmination of a multitude of small steps in innovation, engineering development, and capital investment spanning more than a century.

"Going digital" and getting to the current level of broad digital adoption did not happen overnight. In fact, one of the first steps on this path occurred over 170 years ago, when German scientist Karl Ferdinand Braun observed what he termed the "Point-Contact Rectifier Effect" – one of the first detailed descriptions of the semiconductor junction. At the time, 1847, Braun was working on understanding conductivity in electrolytes and crystals.[1] He observed that putting different materials together, such as a metal and the galena crystal (lead sulfide), resulted in different conductivity depending upon the polarity of the applied voltage. Then, his was just one of many curious observations about materials and their electronic properties that could not be explained by classical physics. It took more than 100 years after Brauns discovery until Bell Labs researchers developed the metal oxide semiconductor field-effect transistor (the MOSFET) in 1959.[2]

Major technological breakthroughs are almost always the result of interdisciplinary efforts enabled by disparate research domains, and the innovation of the MOSFET at Bell Labs is no different. Researchers' development of the MOSFET in 1959 could not have happened without the new electronics and materials science insights provided by quantum physics, a new theory of how light and matter at atomic scales behaved, developed in the 1920s. Quantum

DOI: 10.4324/9780429439278-8

physics explained the effect Braun observed nearly seventy-five years earlier and provided the building blocks for initial electronic engineering to make digital electronic circuits.[3]

Although not realized for decades, the silicon MOSFET had a singularly remarkable ability: the transistor could be miniaturized by more than six orders of magnitude (from millimeters to nanometers) while at the same time, the performance of the individual transistor could improve. Over the next sixty years, the microelectronics industry invested relentlessly in the miniaturization of the transistor, close to doubling the number of transistors in a single chip every two years. This exponential improvement of computing performance over multiple decades also exponentially reduced the cost per transistor, creating more affordable digital technologies, ultimately fueling mass production and mass adoption.

Because of the success of the MOSFET miniaturization, an entire array of other microelectronics technologies was developed and improved using similar methods, including passive components, displays, wireless communications, photovoltaics (solar cells), and batteries. By the turn of the millennium, we could see the path to miniaturizing transistors to atomic dimensions, and now, twenty years later, we are reaching the physical limits of the MOSFET. At this point, improved device performance is fundamentally limited by the number of electrons available to either store memory or switch the transistor circuits on or off. Although we have now reached a definitive physical limit to how small we can reduce MOSFETs, electronic device engineers continue to explore new devices, further integrating semiconductor and optoelectronic capability in increasingly smaller packages, and further reducing the cost per digital function. There is no end in sight.

Sensors allow the digital world to meet the physical world

Continued investments by the microelectronics industry to reduce the size and cost of electronics have also benefited the development of sensors and sensing technologies.[4] Over the same sixty-year period, electronic and optical sensors were miniaturized and their cost similarly reduced. Digital and analog sensor solutions, such as electronic or optical sensors, significantly improved measurement performance compared to their predecessors, which were large, slow, and not very sensitive.

Because of miniaturization, performance improvements, and cost reductions, sensors are now used in all industries and in large volumes by end consumers because they play an essential role in interfacing computers with the physical world. Digital techniques that measure physical, materials, and biochemical processes are increasingly adopted in manufacturing, automation, and consumer digital products. For example, the smartphones, watches, and tablets used by billions of people on our planet include multiple advanced sensor technologies, including megapixel cameras, touchscreen displays, and precision accelerometers, features which are typically taken for granted. These sensors enable the device to physically interact with a range of software applications and to collect data about the physical world.

Optical sensors and imaging systems, for many applications, have the optimum combination of measurement speed, precision, and low cost. Because they typically operate without contacting the sample, optical sensing techniques are applied broadly in a wide range of manufacturing or production processes, but also in medicine, mining, biotechnology, agriculture, and transportation. Optical sensing is particularly useful to rapidly identify materials or quality inspect parts, or to quickly determine the precise distance or position of objects using Light Detection and Ranging or laser interferometry.[5]

Optical sensing methods also accommodate a broad range of measurement scales – from the microscopic to the macroscopic. Optical methods are used to characterize and understand viruses and bacteria (microscopic objects) to plants and rocks (macroscopic objects), to very large-scale regional or global processes including, for example, storm systems or flood extent. The latter is now routinely done from aerial (UAV, drone, or balloon) or space-borne (satellite) imagery platforms. Classifying images, quantifying spatial (area) coverage, or the underlying land or hydrological processes fall under the umbrella of "remote sensing" providing both a detailed and large-scale synoptic view.

Optical methods and remote sensing increasingly provide an indispensable tool set for understanding and managing our environment. Much of our current knowledge about the key environmental processes on our planet (e.g., climate or nutrient, oxygen, and carbon cycles) comes from the optical observations of the earth we have received from satellite imagery. For example, our direct optical observations of phytoplankton (algae) biomass from satellite imagery have enabled us to understand that phytoplankton photosynthesis in global oceans accounts for more than half of the oxygen we breathe. As a result, we also understand that oceans provide a larger biological carbon sink than land and that ocean processes are inextricably coupled to our climate future.

Investments in digital technologies to create positive environmental outcomes

The miniaturization and performance improvements of sensors and other digital technologies have created an opportunity to move beyond traditional concepts of environmental field-based data collection and analysis. Traditionally, field work required costly and time-consuming in situ ground-based surveys. Even as late as the turn of the century, digital measuring equipment deployed in the field was bulky, expensive, and difficult to use. The basic idea of environmental measurement technology of the last thirty years was to develop rugged, field-worthy versions of laboratory measurement equipment. This approach ran straight into the challenges of deploying these technologies in the physical world for prolonged periods including mobile power requirements, temperature extremes, moisture, condensation, and degradation from ultraviolet radiation. The cost of deploying these tools was high, fundamentally limiting access to data collection and limiting the underlying rate of learning.

We now have ready access to the right tools for almost any digital job at the right price. The miniaturization of digital technologies has led to wide availability of low-cost, high-performance computing, sensors, networking, and memory technologies. Internet of Things (IoT) technologies in particular provide robust and scalable solutions for longer-term field deployments operating at low power or autonomously using solar power. With IoT, it is trivial to add sufficient onboard computation to optimize sampling rates, encrypt and reduce data sets prior to transfer to the networked or cloud storage using short range radio or long range cellular or satellite communications.

The performance and low cost of purpose built digital IoT sensor networks together with satellite remote sensing finally provides sufficient information density to begin adequately monitoring a range of land, air, and water environmental factors at low cost. These tools are providing actionable information about what is going on across large geographical areas – entire watersheds, rivers, streams, lakes and reservoirs – with sufficient detail to quickly make informed decisions. Remote sensing combined with continuously running low-cost sensor networks provides unparalleled insight into where problems first arise, where they are going, and how they can be optimally addressed. Additionally, machine learning and artificial intelligence significantly accelerate making sense out of the vastly richer data sets to get to actionable insights quickly.

What we now need is a drastic increase in investments for measurable positive and sustainable environmental outcomes. In particular, a dramatic increase in investments is needed for digital monitoring technologies to track changes of key environmental variables such as water, land, and air pollution. The current spending on impact measurements from environmental remediation or conservation efforts rarely exceeds 1–2 percent of total project costs, despite regulatory burdens that mandate measurement protocols be maintained for multiple years after project completion. This means that for environmental project costs that run in excess of tens to hundreds of million US dollars, less than $100K to $1M is typically allocated to verifying the success of those investments. In contrast, the microelectronics or semiconductor industry invests almost 50 percent of total device manufacturing costs in measurement and process control technologies in order to achieve targeted levels of device improvements every year. With typical fabrication facility equipment costing approximately $24B per factory, companies are spending $1–$2B on tools to measure and control performance improvements are met (per facility!). By comparison, the investment in monitoring protocols for environmental-related projects is a minute fraction.

Increasing investment in measuring and monitoring protocols for environmental projects will increase both the efficiency and impact of those projects. Investing in digital measurement technologies for water, for example, will speed up the feedback loops on the types of land and water remediation, restoration, or management methods that work to most effectively improve environmental outcomes or sustainability. This improved feedback will in turn help identify the costs of inaction and determine which environmental solutions reach targeted outcomes at lowest cost and in the shortest time period. Once digital measurements at watershed scale are paired up with technologies that result in environmental improvements, significant additional capital will be unlocked to achieve even more ambitious goals.

Increased investment is urgently needed now. While increased awareness and desire to improve environmental outcomes is growing, which is helping to close investment gaps between IoT and environmental technology. The accelerating environmental degradation and impact on our climate – most acutely on the sustainability of our water and food systems – are already resulting in an escalating need for action. Communities, governments, and businesses are increasingly facing direct operational and financial impacts from increasingly severe natural disasters. This comes in the form of too much or too little water or uncertainty about access to clean water for sanitation, drinking, irrigation, or industrial production. The breadth and depth of recent natural disasters including devastating wildfires in Australia and the western United States, floods in the southern United States, and persistent droughts in the Colorado basin or acute intermittent ones in South Africa are increasing the urgency for action to mitigate climate change and implement sustainable solutions. Private and public sector investment must focus on narrowing the investment gap before the point of no return.

Optical remote sensing technologies have enabled the digital world to interface with, and make sense of, our physical world. This increased understanding enables us to build large-scale digital models that can analyze climate change, refine environmental scenarios, enable us to act to mitigate or manage local environmental impacts, and start understanding how to manage key physical processes. In many ways, the broad adoption of sensing and digital technologies has allowed us to learn more about our physical world and to move faster toward understanding our place in the universe and more pressingly on our planet. Digital technologies such as these may be our only hope to speed up innovation of solutions for managing our natural world for sustainable value creation, for individuals, communities, and businesses across geographies and scales.

We no longer have the technology (or cost of technology) barriers to address the growing environmental problems we face around water and food production. "Going digital" in the physical world – applying our broad levels of digital technology to better manage our physical environment and improve water, soil health, and other environmental sectors – is already demonstrated in practice and is bearing fruit. Increasingly, a range of computational applications, including machine learning and artificial intelligence, are guiding drinking water, wastewater, and stormwater processes and management decisions. Drinking and wastewater utilities are optimizing treatment processes in real time, NGOs are tracking their environmental restoration impacts, and governments are evaluating policy or infrastructure investment decisions. Near-term meteorological forecasts and climate models are improving, and businesses are better managing operational, regulatory, and reputational risks and are able to quantify the effectiveness of different mitigation approaches in the face of rising uncertainty. If we can successfully scale these efforts, these environmental investment-to-outcome feedback loops have the potential to truly revolutionize how we manage the key ingredients of life on this planet. Digital technologies have the potential to yield and accelerate positive and sustainable environmental outcomes.

Notes

1 The point-contact rectifier effect describes the electrical performance of a junction between metals and certain crystals, which do not conduct current equally when voltage polarity is switched. Braun was specifically interested in electrolysis.
2 MOSFET, the now-standard digital transistor, powers memory devices and computation circuitry in nearly all microprocessors commonly used today.
3 Braun's observation, thanks to the explanations of *quantum* physics, became the basic building block of a digital electronic switch – a semiconductor diode or junction. Additionally, photoconductivity and the photoelectric effect, were some of the many phenomena that, once understood, would enable the development of switching, sensing and photovoltaic (solar) power technologies.
4 Sensors are typically electromagnetic or optoelectronic devices that measure or record physical (chemical, biological or geological) properties.
5 LIDAR and laser-based interferometry are remote sensing techniques which enable the accurate detection of the location and velocity of a reflective target; both are being used in precision robotics and autonomous vehicles for computer detection of the physical environment.

Index

Note: Page numbers in *italics* indicate a figure on the corresponding page.

Printed in the United States
by Baker & Taylor Publisher Services